CW01511566

THE COSTS AND BENEFITS OF MANAGING WILD GEESE IN SCOTLAND

Douglas MacMillan, Mike Daw,
Dilly Daw, Lorna Phillip and Ian Patterson
(University of Aberdeen)

Nick Hanley (University of Glasgow)

Julie-Ann Gustanski (University of Edinburgh)

Robert Wright (University of Stirling)

Funded by Scottish Executive Agricultural and
Biological Research Group Flexible Fund

Scottish Executive Central Research Unit
2001

Further copies of this report are available priced £5.00. Cheques should be made payable to The Stationery Office Ltd and addressed to:

> The Stationery Office Bookshop
> 71 Lothian Road
> Edinburgh
> EH3 9AZ
>
> Tel: 0131-228-4181
> Fax: 0131-622 7017

The views expressed in this report are those of the researchers and do not necessarily represent those of the Department or Scottish Ministers.

© Crown Copyright 2001
Limited extracts from the text may be produced provided the source is acknowledged. For more extensive reproduction, please write to the Chief Research Officer at the Central Research Unit, Saughton House, Broomhouse Drive, Edinburgh EH11 3XA

CONTENTS

TECHNICAL REPORTS
(Only available on the CRU website: www.scotland.gov.uk/cru)

REPORT A	**VIEWS ACROSS SCOTLAND ABOUT THE CONSERVATION AND MANAGEMENT OF WILD GEESE IN SCOTLAND**
REPORT B	**WILLINGNESS TO PAY FOR THE CONSERVATION AND MANAGEMENT OF WILD GEESE IN SCOTLAND**
REPORT C	**THE COST TO SCOTTISH AGRICULTURE OF WILD GEESE: ISLAY AND LOCH OF STRATHBEG CASE STUDIES**

LIST OF ABBREVIATIONS

CE   Choice Experiments

CV   Contingent Valuation

ESA   Environmentally Sensitive Area

GMS   Goose Management Scheme

IVGMS   Islay Voluntary Goose Management Scheme

MS   Market Stall

SERAD   Scottish Executive Rural Affairs Department

SNH   Scottish Natural Heritage

SSSI   Site of Special Scientific Interest

WTP   Willingness To Pay

WTA   Willingness To Accept

EXECUTIVE SUMMARY

Scotland is an important destination for migratory geese species such as the internationally-protected populations of Greenland Barnacle and Greenland White-fronted geese on Islay and the more numerous and widely dispersed populations of Pink-foot and Greylag geese in eastern Scotland.

Over the past thirty years the number of geese over-wintering in Scotland has increased rapidly due to changes in agricultural practice, special management actions by the RSPB and other conservation groups, and legislative protection. For example, the population of the Greenland Whitefronted Goose has doubled in the last 20 years.

The rising goose population has brought goose conservation into conflict with farmers, due to the damage geese can inflict on agricultural crops. On Islay, Scottish Natural Heritage (SNH) has paid farmers compensation for goose damage and similar goose management schemes are being planned for other areas.

As the cost of compensation and crop damage will increase if the goose population expands further, there is some concern that the costs to government may become excessive. Are future increases in the goose population therefore desirable? A key recommendation of the National Goose Forum was that an economic study of the costs and benefits of Scotland's goose population should be undertaken to help answer this question.

The main aim of this project was to provide comprehensive qualitative and quantitative information about the economic costs and benefits of wild geese in Scotland. The research focuses on two case study areas where conflicts have arisen between geese and agriculture: Islay and Loch of Strathbeg in north-east Scotland.

Specific objectives were to:

1.   *Describe the wider views and opinions of the general public, local residents, visitors, and tourist businesses about the conservation and management of wild geese*

2.   *Value in monetary terms the non-market benefits of wild goose conservation to the general public, local residents and visitors.*

3.   *Estimate the costs of goose damage to agriculture*

The research therefore had both a national and a local focus. At the national level the research involved a representative sample of the general public, while in the two case study areas, the research targeted four important stakeholder groups: residents, tourist visitors, tourist business operators, and farmers.

The research effort involved a range of approaches:

1.   Qualitative research based around group discussions, where the perceptions and awareness of the different stakeholder groups regarding wild goose conservation in Scotland could be explored in-depth (Chapter 4).

2.   Quantitative research involving survey questionnaires to estimate the value placed on policies to conserve wild geese by individual households across Scotland, plus visitors and residents in the two case study areas. This was achieved by asking people what they would be willing to pay in additional taxation for the conservation of wild geese (Chapter 5).

3.   A survey of farmers on Islay and around Loch of Strathbeg who have been affected by the presence of large numbers of wild geese on their land in order to estimate the costs of wild geese to agriculture (Chapter 6).

The main findings were:

General Views

- *Awareness about geese......*
  Local residents in Strathbeg and on Islay felt that the "presence" of wild geese added something indirectly to their life in the region, but for some this impact was positive, while for others it was negative. Compared to local residents, the general public was less aware of, and interested in, issues involved with wild geese (See Table 4.1, pages 19-22).

- *The importance of conserving wild geese...*
  Across all groups goose conservation was considered to be of low priority, with only "re-introducing beavers" coming lower in a list of 6 different nature conservation policies (see pages 32 and 37). However, most people felt that endangered species of geese such as the Greenland White-fronted goose should have more protection than non-endangered species such as the Pink-footed goose.

- *Control of geese......*
  Most people understood the need to protect crops and compensate farmers, but took issue with any control mechanism that caused direct harm to geese (e.g. shooting). However, local residents were less opposed to shooting as a means of controlling geese than either the general public or visitors.

Geese and Tourism

Although it was not within the remit of the study to quantify the contribution wild geese make to the local economy some attempt to gauge the importance of geese to tourism was attempted through discussions with local tourism representatives. Key findings were:

- *Geese and the local economy.....*
  Geese did play a role in the tourist economy of both case study areas. On Islay, geese are one of many attractions for visitors, while goose shooting helps sustain the economy of the Strathbeg area in the winter months.

- *Do farmers benefit?…..*
  It was generally agreed that farmers did not benefit directly from goose related tourism, apart from a few farmers at Strathbeg who received payments for providing shooting access (see Table 4.2, pages 24-26).

Benefits of Goose Conservation

Monetary valuation of the benefits of wild geese was attempted by asking people for their 'willingness to pay' for a range of conservation policy options for wild geese.

- ***What kind of management policy do people want?……***
  The survey found that different attributes of goose conservation policy were valued differently by the various groups. For instance, the general public and visitors were willing to pay between £7-£19 per household per year for a management policy that did not involve shooting. Residents, on the other hand, were not willing to pay anything for this type of policy (see Table 5.4, page 39).

- ***Do the public value all geese species or only endangered species?…..***
  The qualitative and quantitative research strongly suggests that all groups favoured conservation policies that target endangered species. Statistical analysis of the willingness to pay results showed that the general public were not prepared to pay significantly higher levels of additional taxation for polices that extended conservation measures to non-endangered species (see page Table 5.2, page 34). Both visitors and local residents were actually prepared to pay **more** for a policy that included endangered species only, than one that included all species (see Table 5.4, page 39).

- ***Do the public value further increases in goose population?...***
  There was some evidence that most people would be willing to pay for policies that encouraged a small increase (10%) in endangered species, but would not pay for bigger increases in goose numbers. For example, local residents said they would be willing to pay £13 per household on average to **avoid** a 50% increase in the goose population (see Table 5.4, page 39).

- *Aggregate willingness to pay for different policy options?...*
  The total annual willingness to pay of the Scottish population for some alternative goose management policies was as follows (see page 47):

For a policy that does not rely on shooting geese:	**£5.2 million**
For a policy that prevents a 10% fall in endangered species	**£6.8 million**
For a policy that achieves a 10% rise in endangered species	**£10.2 million**

- *Validation of willingness to pay surveys* …
  As the estimates of willingness to pay given above were hypothetical in the sense that they are a response to a survey question and did not involve an actual payment, it is important to verify that they reflect accurately people's preferences. To do this we investigated the extent to which individual WTP responses were consistent with prior expectations (that is, made sense!) and we found WTP was positively correlated with income and how important the respondent thinks protecting wildlife is relative to

other goals of rural policy. Also a comparison of mean WTP estimates derived from two independent samples of the general public using two different techniques were not significantly different (in one survey WTP for stop shooting was £9.23, while in the other it was £8.32). These are re-assuring findings (see pages 39-41).

Agricultural Costs of Geese

The following estimates of the costs of wild geese to agriculture are based on a detailed survey of 18 farmers on Islay and 15 farmers in the Strathbeg area. In order to obtain total costs it was necessary to scale up the estimates to all farms in each case study area. While this introduces some sampling error we believe that the figures given below are accurate estimates of goose damage at both locations for the 1999/2000 season. However it should be remembered that these estimates are for one year only, and that damage estimates will vary from year to year due to variation in the type of crops grown, weather, goose densities and prices.

- *The total and average cost of goose damage......*
  The estimated total costs of goose damage on Islay and Strathbeg in 1999/2000 were £560, 000 and £220, 000 respectively. The higher costs on Islay reflect the higher densities of geese found there. The average cost of damage per goose was around £13 at both locations. These results are consistent with estimates from earlier studies (see Table 6.5, page 60).

- *The marginal costs of goose damage......*
  At both locations, rising goose numbers lead to greater economic costs to farmers, but the survey suggests for Islay that the cost per goose decline rapidly as numbers rise to high levels (see Figure 6.2, page 57). On Islay, where the geese are concentrated on a relatively small area this may be due to competition among the geese, and the fact that the Barnacle goose, which is smaller than the White-fronted species (and hence eats less), makes up a higher proportion of the total population at high densities. At Strathbeg, and other locations where geese can easily move to other areas, there is less competition and the decline in the marginal cost of damage is less pronounced (see Table 6.4, page 59).

Recommendations for Further Research

- No attempt was made to estimate the non-market 'costs' of geese. Instead of being prepared to pay something to further protect geese, people who have negative views about geese would wish to be compensated. Incorporating willingness to accept compensation as a measure of the non-market costs of geese may lower the overall benefits of policies to protect wild geese.

- Further investigation and quantification of the benefits of geese to tourism is merited.

- It is also recommended that further survey work is undertaken to investigate how the costs of geese damage vary on an annual basis, particularly in response to changes in the goose population. The current study has only examined the marginal cost of goose damage across a range of goose densities on different farms – in order to predict the marginal cost of further increases in goose numbers data from a number of years would have to be collected.

- More regular and detailed goose counts throughout the country would enable a more detailed investigation of the total cost of wild geese in Scotland.

CHAPTER ONE   INTRODUCTION

Summary

Wild geese generate a range of costs and benefits to society. On the one hand, geese can cause significant damage to farmers' crops especially around bird reserves where the geese congregate. On the other, people value geese as a recreational resource (e.g. for shooting or watching), or simply as wild birds that ought to be conserved for future generations.

The overall aim of this project is to provide comprehensive qualitative and quantitative information about the economic costs and benefits of wild geese in Scotland. The research focuses on two case study areas where conflicts have arisen between geese and agriculture: Islay and Loch of Strathbeg.

Specific objectives are to:

- Describe the wider views and opinions of the general public, local residents, visitors, and tourist businesses about the conservation and management of wild geese.

- Value in monetary terms the non-market benefits of wild goose conservation to the general public, local residents and visitors.

- Estimate the costs of goose damage to agriculture.

1.1 Background

1.1.1 Scotland is an important destination for migratory geese species. Internationally-protected populations of Greenland Barnacle and Greenland White-fronted geese congregate on Islay every year and the more numerous Pink-foot and Greylag geese are dispersed widely in eastern Scotland (Figure 1.1).

1.1.2 Over the past thirty years the number of geese over-wintering in Scotland has increased rapidly due to changes in agricultural practice, special management actions by the RSPB and other conservation groups, and legislative protection. For example, the population of the Greenland Whitefronted Goose has doubled in the last 20 years.

1.1.3 The rising goose population has brought goose conservation into conflict with farmers, due to the damage geese can inflict on agricultural crops. During the winter and early spring months wild geese depend to a great extent on farmland for their food supply. Grazing by geese, especially in areas where goose numbers are highly concentrated, can cause damage to spring-sown cereals and grass, delay turn-out of stock, and can cause problems with soil puddling and compaction.

1.1.4 The range of measures for controlling damage by geese available to farmers is restricted. Shooting endangered species is strictly controlled and other preventative measures such as scaring are often ineffective. In order to fulfil international conservation obligations and to help farmers in areas badly affected by wild geese, the government intends to introduce local goose management schemes. These schemes will provide special feeding and buffer areas for geese on farmland and, in return, farmers will receive compensation. The total cost of implementing local management plans is not known yet but it is likely to be substantial. A previous scheme on Islay, for example, cost the government over £400 000 in 1998-99.

1.1.5   As the cost of compensation and crop damage will continue to increase if the goose population expands, there is some concern that the costs to government may become excessive. Are further increases in the goose population therefore desirable?   A key recommendation of the National Goose Forum was that an economic study of the costs and benefits of Scotland's goose population should be undertaken.

1.1.6   Wild geese can generate a range of benefits for different groups. People may positively value wild geese populations for recreational reasons such as shooting and viewing (use values), and from the pleasure of knowing they exist (non-use value). In addition visitor expenditure from both birdwatchers and goose hunters, provide employment and enhance the profits of local businesses such as hotels, restaurants, and shops.

Figure 1.1 Distribution of Main Species of Wintering Geese in Scotland

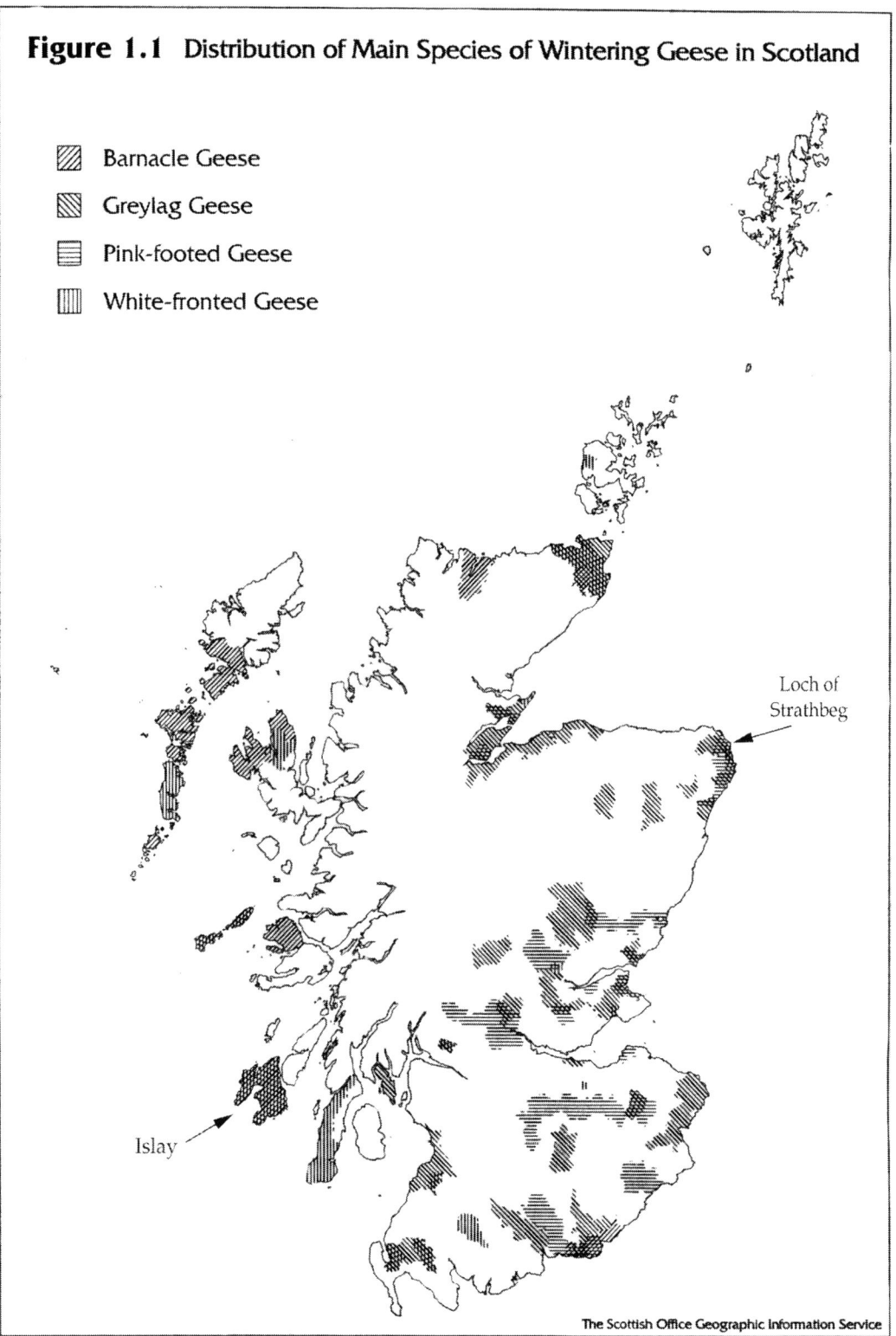

- ▨ Barnacle Geese
- ▧ Greylag Geese
- ▤ Pink-footed Geese
- ▥ White-fronted Geese

Loch of
Strathbeg

Islay

The Scottish Office Geographic Information Service

3

1.1.7 In the case of shooting, farmers and other landowners can sell the rights to shoot and hence directly benefit from the presence of wild geese. However, in the case of other recreational activities (bird-watching) and non-use values, markets are difficult to create. For example, it would be prohibitively expensive to charge people for watching geese and impossible to exclude them from caring about geese.

1.1.8 Fortunately there is an alternative way to approach valuing benefits through survey-based valuation methods such as contingent valuation (CV) and choice experiments (CE). These methods rely on creating a realistic, though hypothetical market, where people can express their willingness to pay (WTP) for wild goose conservation and management. (In economics, how much we want something is estimated by our willingness to pay).

1.2 Objectives

1.2.1 The overall aim of this project is to provide comprehensive qualitative and quantitative information about the economic costs and benefits of wild geese in Scotland to farmers, local residents, tourists, and the general public.

1.2.2 The research focuses on two case study areas where conflicts have arisen between geese and agriculture: Islay and Loch of Strathbeg. Specific objectives are to:

- Describe the wider views and opinions of the general public, local residents and visitors about the conservation and management of wild geese in Scotland.

- Assess the importance of wild geese to tourism.

- Estimate the non-market benefits of wild goose conservation through WTP surveys of the general public, local residents, and visitors.

- Estimate the costs of goose damage to agriculture.

1.3 Report Structure

1.3.1 The report is divided into seven chapters. The next chapter (Chapter 2) describes the research methods employed in the course of the study. Chapter 3 provides an overview of the main results. Chapters 4, 5, and 6 report on the main findings of the research: Chapter 4 describes the qualitative research undertaken to explore the views and opinions of various groups about wild geese conservation in Scotland; Chapter 5 describes the results of the WTP surveys, and Chapter 6 reports on the results of research into agricultural costs. Conclusions and recommendations for future research are presented in Chapter 7.

1.3.2 There are also three technical reports which are bound separately. These reports provide a detailed account of the methods and results of the qualitative research (Report A), the WTP surveys (Report B), and Agricultural Costs (Report C).

CHAPTER TWO   STUDY METHODS

Summary

In order to provide a comprehensive investigation of the issues surrounding wild geese in Scotland an integrated approach, involving a range of research methods was adopted, including focus groups, survey questionnaires, and one-to-one interviews

The research effort had four distinct phases:

(1)   Information gathering;
(2)   Qualitative research involving focus groups;
(3)   Quantitative research involving willingness to pay surveys;
(4)   an in-depth assessment of the costs of wild geese to agriculture.

The research had both a national and a local focus. At the national level the research involved members of the general public. In the case study areas the research was targeted at four local stakeholder groups: local residents, tourist visitors, local tourist businesses, and local farmers.

2.1   Introduction

2.1.1   A range of qualitative and quantitative techniques were used to provide a comprehensive picture of the costs and benefits of wild geese. The research focused on both the national picture with surveys of the general public and the local situation where research was conducted with different four stakeholder groups: local residents, local tourist businesses, tourist visitors and local farmers. Two sites were picked for detailed local study:

1. Islay: where a compensation scheme exists for the conservation of populations of internationally protected Greenland Barnacle and Greenland White-fronted Geese.
2. Loch of Strathbeg: where quarry species such as the Greylag and Pink-footed Geese are found.

The location of these case study sites is shown in Figure 1.1.

2.1.2   The research effort had four distinct phases: (1) information gathering; (2) qualitative research involving focus groups; (3) quantitative research involving willingness to pay surveys; and (4) an in-depth assessment of the costs of wild geese to agriculture. Each phase is briefly described below.

2.2   Information Gathering

2.2.1   The information gathering stage involved reviewing key reports and a series of meetings with individuals who could provide expert briefing on policy and scientific issues such as current goose conservation measures, goose feeding and damage, and population modelling. Information sheets, containing both textual and visual information about conservation and management of the four main migratory species were developed for the focus group stage. (An example of one of the Information Sheets used is given on the next page).

2.3   Qualitative Research

2.3.1   In total 12 focus groups, each involving between 5-8 people were planned to cover each of the main study target groups (Table 2.1). These groups provided a forum for exploring how people conceptualise and view the issues surrounding wild geese. Specific subjects addressed included ethical and cultural attitudes towards geese and their management, the relative importance of geese in terms of conservation, and the problem of agricultural damage. Methodological matters such as the method of payment (taxation or charitable donation) and other details relative to the willingness to pay survey also featured in the 'topic guide' for the focus groups.

Table 2.1	Location and target group for Focus Groups				
	Participant Groups				
Location	Local Residents	Farmers	Tourist Trade	Visitors	General Public
Islay	1	1	1	1*	-
L. of Strathbeg	1	1	1	1	-
Stirling	-	-	-	-	2
Edinburgh	-	-	-	-	2

* This group did not meet due to low visitor numbers on Islay in late February.

2.4   Willingness to Pay Surveys

2.4.1   Monetary valuation of the benefits of wild goose conservation benefits was achieved through the application of two survey-based valuation techniques: choice experiments (CE) and contingent valuation (CV). CE involve people making choices between alternative policy options that are described in terms of specially selected attributes. CE is becoming increasingly popular because it can provide a statistically efficient means of estimating WTP for marginal changes in a range of attributes that are of policy interest.

2.4.2   CV is the most established non-market valuation method and involves asking people for their WTP for a specified environmental policy option in a hypothetical market context. CV was employed to provide WTP estimates for four distinct goose management options of policy interest based which could be used to help validate the marginal WTP estimates derived from the CE exercise.

2.4.3   Following an intensive period of pre-testing and piloting the valuation surveys were implemented by System Three, a market research company. Interviews were carried out with individuals selected using the sampling frame described in Table 2.2, that was agreed with SERAD.

2.4.4   Individuals within each category were selected using a quota sampling approach based on age, income, and gender. Islay and Strathbeg resident interviews took place in April and May 2000, with interviews involving the general public taking place between June and

July 2000 throughout Scotland. These interviews took place in the home. Visitors were interviewed at a range of local tourist attractions including the local RSPB reserve, and hence could include both resident and non-resident visitors. Some Islay visitor interviews also took place on the ferry service to and from the mainland. Visitor interviews took place in late April/early May and in October in order to interview people who may have been visiting the case-study to watch the geese arrive or depart. (As interviews did not take place in the winter the visitor sample does not include goose shooters who tend to visit Strathbeg between November and January).

Table 2.2	Number of CV and CE interviews		
	Location		
Target Population	Islay	L. Strathbeg	Scotland
Local Residents (CE)	200	200	-
Visitors (CE)	200	200	-
General Public (CE)	-	-	400
General Public (CV)	-	-	400

2.4.5 Due to the hypothetical nature of WTP valuation surveys it is important that considerable effort is placed on validation. In addition to standard internal checks this study introduced three further validation tests.

- Convergent validity: This refers to assessing the extent to which alternative valuation methods generate similar values. In this study this was achieved by comparing equivalent WTP estimates for both CE and CV surveys.
- Feedback Sessions: in small groups of 4-8, individuals who participated in the main survey were asked to critically discuss the valuation exercise they were involved in and to assess the reliability and accuracy of their responses.
- A group-based valuation technique called the CV Market Stall was also implemented. The Market Stall approach involves participants in two group sessions one week apart and hence gives individuals greater opportunity to consider and discuss their WTP than is possible with more conventional survey-based approaches. In total 52 participants, in 8 groups from around Scotland (Nairn, Aberdeen, East Kilbride, and Dumfries) were recruited for this part of the project.

2.5   Agricultural Costs

2.5.1 This phase of the research involved a broad-ranging literature review, two Focus Groups involving 6-8 farmers from Islay and from Strathbeg, and a detailed farm-level survey questionnaire about goose damage. The latter was achieved through individual visits to a representative sample of farms affected by wild geese in each case study area.

CHAPTER THREE   OVERVIEW OF RESULTS

Summary

This chapter provides a brief overview of the main results relating to the:

- Qualitative research on attitudes toward the conservation of wild geese.

- Role geese play in the local tourist economy.

- WTP surveys to estimate the non-market benefits.

- Agricultural costs of wild geese.

Public opinion

- *Awareness about geese...*
  Local residents in Strathbeg and on Islay felt that the "presence" of wild geese added something indirectly to their life in the region, but for some this impact was positive, while for others it was negative. Compared to local residents, the general public was less aware of and interested in issues involved with wild geese.

- *The importance of wild geese to people...*
  Across all stakeholder groups goose conservation was considered to be of low priority, with only "re-introducing beavers" coming lower in a list of 6 different nature conservation policies. However, most people felt that endangered species of geese such as the Greenland White-fronted goose should have considerably more protection than non-endangered species such as the Pink-footed goose.

- *Control of geese...*
  Most people understood the need to protect crops and compensate farmers, but took issue with any control mechanism that caused direct harm to geese (e.g shooting). However, local residents were less opposed to shooting as a means of controlling geese than either the general public or visitors.

Geese and tourism

- Geese play a role in the tourist economy of both case study areas. On Islay, geese are one of many attractions for visitors, while goose shooting helps sustain the economy of the Strathbeg area in the winter months. However, it was generally agreed that farmers did not benefit directly from goose related tourism, apart from a few farmers at Strathbeg who received payments for providing shooting access.

Benefits of goose conservation

- *What kind of management policy do people want?......*
  The survey found that different attributes of goose conservation policy were valued differently by the various groups. For instance, the general public and visitors to both areas were willing to pay between £9, £6 and £19 per household per year (over 10 years) respectively for a management policy that would not involve shooting. Residents, on the other hand, were not willing to pay anything for this type of policy.

- *Do the public value all geese species or only endangered species?.....*
  The qualitative and quantitative research strongly suggests that all groups favoured conservation policies that target endangered species. Statistical analysis of the willingness to pay results showed that the general public were not prepared to pay significantly higher levels of additional taxation for polices that extended conservation measures to non-endangered species. Visitors and local residents were actually prepared to pay **more** for a policy that included endangered species only, than one that included all species.

- *Do the public value further increases in goose population?...*
  There was some evidence that most people would be willing to pay for policies that encouraged a small increase (10%) in endangered species, but would not support or pay for bigger increases in goose numbers.

- *Aggregate willingness to pay for different policy options?...*
  The total annual willingness to pay of the Scottish population for a goose management option that does not allow shooting geese and for one that achieved a 10% rise in endangered species was £5.2 million and £10.2 million respectively.

- *Validation of willingness to pay surveys ...*
  As the above estimates of willingness to pay (WTP) are based on a response to a survey questionnaire and do not involve an actual payment it is important to verify that they are an accurate reflection of people's preferences. To do this we investigated the extent to which individual WTP responses were consistent with prior expectations (that is, made sense!). We also found that WTP was positively correlated with income and how important the respondent thinks protecting wildlife is relative to other goals of rural policy. A comparison of mean WTP estimates from the CE and CV samples of the general public for a policy option that prevented shooting were not significantly different. (In one survey WTP for stop shooting was £9.23, while in the other it was £8.32). These are re-assuring findings and gives us some confidence that these hypothetical estimates of WTP are a reliable reflection of the value the general public place on wild goose conservation.

Agricultural costs of geese

- *The total and average cost of goose damage to agriculture...*
  The estimated total cost of goose damage on Islay and Strathbeg in 1999/2000 was £560 000 and £220 000 respectively. The higher costs on Islay reflect the higher densities of geese found there. The average cost of damage per goose (which takes

account of variation in density) was around £13 at both locations. These results are consistent with cost estimates from previous studies.

- ***The marginal costs of goose damage......***
  At both locations, rising goose numbers lead to greater economic costs to farmers, but the survey suggests for Islay that the cost per goose decline rapidly as numbers rise to high levels. On Islay, where the geese are concentrated on a relatively small island this may be due to strong competition among the geese, and the fact that the Barnacle goose, which is smaller than the White-fronted species (and hence eats less), makes up a higher proportion of the total population at high densities. At Strathbeg, and other locations where geese can easily move to other areas the decline in the marginal cost of damage is less pronounced.

CHAPTER FOUR   QUALITATIVE RESEARCH

Summary

The qualitative research had two principal aims:

- To provide a rich and comprehensive insight into opinions and views about the conservation and management of wild geese populations in Scotland

- Contribute to the design of the willingness to pay studies

The research on public opinion focused on three groups: local residents and visitors in the case study areas and the general public in Scotland.

In order to investigate the impact of geese on the local economy, a tourism focus group was held in each case study area. Participants included hotelliers, caterers, shop owners, and managers of tourism facilities. (The qualitative research involving local farmers is described in Chapter 6.)

The chapter describes the principal results of this research in summary form, with more detailed information and analysis available in Technical Report A.

Key Findings

Local Residents, Visitors, & General Public....

- *What criteria should influence government conservation priorities for wildlife?*
  It was generally thought that the most influential factors were: endangered status, current population level, breeding characteristics and migration patterns, and protected status (national, EU, or international).

- *How important are geese to you?*
  Most of the general public exhibited a general indifference towards wild geese as they make little impact on their day-to-day lives. Participants in the Strathbeg and Islay groups felt that the "presence" of wild geese added something indirectly to their life. For some this impact was positive, while for others it was negative.

- *Do you want more geese?*
  Increases in the rarer species such as the Greenland Barnacle and White-fronted were generally desirable, but local residents tended to be more in favour of controlling the expansion in the goose population.

- *What about agricultural damage?*
  People were concerned about the impact geese have on farming, and most understood the need to protect crops. However, most people, except Strathbeg residents, took issue with any control mechanism that caused direct harm to the geese (e.g shooting) Compensation payments to farmers were widely regarded as a

practical solution, although other options such as changes to farming systems should be considered first.

Local Tourism representatives...

- *Importance of tourism to the area?*
  Tourism is relatively more important to the local economy of Islay than to the Loch of Strathbeg area.

- *How important a role do geese play in attracting tourists?*
  Geese play an important role in the tourist economy of both places: Goose-watching is of some importance to bird-watchers on Islay, whereas goose shooting is more important around the Loch of Strathbeg.

- *Would fluctuations in the number of geese affect tourism?*
  Moderate fluctuations in the size of the goose population were not thought likely to affect tourism in either area.

- *Views on shooting?*
  Commercial shooting is not considered to be an option at the moment on Islay, but there was widespread agreement that the local economy of Strathbeg would benefit from extending the open season to February.

4.1 Introduction

4.1.1 The qualitative research aimed to provide a rich and comprehensive investigation of opinions and views about the conservation and management of wild geese populations in Scotland. The research focused on three groups targeted in the WTP study: Islay and Strathbeg residents and visitors, and the general public. Two additional groups involving local people involved in tourism were also convened to inform the researchers about the impacts of wild geese on tourism.

4.1.2 In relation to the subsequent WTP study these groups served as a first step in understanding the importance and value placed on four migratory geese species (Greenland Barnacle, Greenland White-fronted, Pink-footed and Icelandic Greylag) relative to larger and more complicated issues of species conservation and other environmental and social issues. They also provide insight into how the general populace across Scotland view the damage caused to farm crops by geese, the need for compensation payments, and the nature of conservation measures. The research therefore had a central role both in terms of:

- designing the subsequent valuation studies
- informing future government policy

4.1.3 The tourism groups involved members of the tourist trade in Islay and Strathbeg and focused on understanding the extent to which geese attract visitors to the local area and the importance of expenditure associated with these visits to the local economy. This research is described at the end of the chapter.

4.2   Local Residents, Visitors, & General Public

4.2.1   The focus groups were conducted in four geographic locations across Scotland in late February/early March 2000 (see Table 2.1). The groups lasted for about 2 hours and were structured to allow participants' views about geese conservation and management to unfold as the discussion progressed. Initial discussions were general in nature and concerned topics such as environmental responsibility and the protection of wildlife. Subsequent topics were more specific to wild geese and future management options.

4.2.2   To aid discussion participants were provided with sheets that provided general information about each of the four main migratory species. The information covered topics such as geographic location, numbers and distribution, population changes over time, vulnerability to extinction, legal status, and agricultural damage. (An example of an Information sheet is given on page cards is given on page 8).

4.2.3   The discussions were analysed with the results are fully documented in Technical Report A, together with all material prepared for the Focus Groups. The summary analysis presented in Table 4.1 offers key insights on popular attitudes toward goose conservation across the three target groups.

4.2.4   As views on most topics about geese varied within groups, as much as across groups, it was decided to summarise the discussions by providing a digest of 'converging' and 'diverging' views, rather than the views of each group on each topic. Where a divergent view is given by a particular stakeholder group this group is identified.

4.2.5   The Focus Groups provided a number of insights of relevance to both policy–makers and to the design of the WTP survey.

- Compared to local residents and visitors, the general public was less aware of and interested in issues involved with wild geese and overall tended to have less polarised opinions.

- Most of the general public exhibited a general indifference towards wild geese as they make little impact on their day-to-day lives. Most participants in the Strathbeg and Islay groups stated that the "presence" of the wild geese added something indirectly to their life in the region, but for some this impact was positive, while for others it was negative.

- Most important criteria determining priorities for conservation were endangered status, current population level, breeding characteristics, migration patterns, and protected status (national, EU, or international).

- Population increases in the rarer species such as the Greenland Barnacle and White-fronted were considered generally desirable, but local residents tended to be more in favour of controlling the expansion in geese numbers

- Most groups understood the need to protect crops, but took issue with any control mechanism that caused direct harm to the geese (e.g. shooting). Habitat management to control the food supply was thought to be a better approach.

- Many people felt that compensation payments to farmers were fair but that the need for compensation should be minimised if possible by changing agricultural practices.

- Common to nearly every participant across focus groups was the lack of willingness to pay in terms of any additional taxes to support goose conservation measures.

- In relation to the design of the WTP surveys it was clear from the Focus Group discussions that the issue of compensation for farmers using tax-payers money would have to be carefully explained.

- In terms of policy scenarios for the Choice Experiment the key attributes appeared to be conservation status, method of control, and future changes in goose numbers.

4.3 Tourism Representatives

4.3.1   In many parts of Scotland during the period October to May migratory wild geese are an important attraction for birdwatchers, for wildfowl shooters and for the general tourist. While on holiday, these visitors spend significant sums of money on food, accommodation and other goods and services which helps to support the local tourist industry during the quieter winter period.

4.3.2   There have been two studies that have specifically examined the economic impact of wild geese on tourism in Scotland. Marston (1998) interviewed 53 local tourist business proprietors on Islay, and a report by RSPB/BASC (1998) used data from visitor surveys to estimate the total number of jobs and income generated by goose-related visitor expenditure in Scotland.

4.3.3   Key findings of the RSPB/BASC report (which was prepared for the National Goose Forum) were:

- Goose shooters spent over £2 million in Scotland in 1997/98
- Over 90% of shooters visited the area specifically to shoot geese
- Goose watchers spend an estimated £3 million in local economies in Scotland of which 50% was assumed to be attributable to the presence of geese
- Key sites for geese watching are Vane Farm (Tayside), Caerlaverock, Mersehead, Ken and Dee marches (Dumfries and Galloway); Islay (Argyll), Nigg and Udale Bays (Cromarty) and Loch of Strathbeg (Grampian).
- More than 100 FTE jobs are supported by visitor expenditure associated with geese watching or shooting, with just over half of this employment generated by goose shooting and wildfowling (58%), with the remainder attributable to goose watching.

4.3.4 The overall conclusion of both studies is that goose watching and goose shooting contribute significantly to the rural economy in localised areas especially during the period October to March/April when other visitors have left.

Table 4.1: Summary of Local Resident, Visitor and General Public Focus Group Discussions

Topic: Protection of the environment

Question….	Converging Views	Diverging Views
An important issue?	Focus group participants across Scotland were unanimous in that they believed protecting the environment was important, but not as important as some other things. A typical comment was: *"Yes, it is important to protect our environment, but relative to other things the government should be concerned with, I would put it second or third, but not the first important issue for government to concentrate on.*	
Who is responsible?	Both the government and the individual shared responsibility *"we are all responsible to ensure the environment is protected for the future" "the responsibility to ensure the environment is protected belongs with everyone with government leading the way"*	Particular to the Islay resident group, but mentioned by participants of other focus groups, was the focus on the RSPB as having a key role to play.

Topic: Priorities for Species Protection

Which species?	All species should have some level of protection but most threatened with extinction should have greatest protection	The Islay residents focus group were concerned that some protected species were "out of control" (e.g geese).
Criteria?	The criteria for determining priorities mentioned across groups were **endangered status / threat of extinction**; current **population level**; **breeding characteristics** and **migration patterns**; restoration of **balance**; **protected status** (national, EU, or international); **relative importance** of one species compared to another; and **indigenous** and/or **geographic concentrations** of particular species. . Most participants squarely saddled the government with *the "responsibility for sorting it out".*	Unique to the Loch of Strathbeg visitors group was the concern over the protection of habitat. Participants in the group viewed habitat as the single most critical factor.

Topic: Personal knowledge/impacts of wild geese

Knowledge/ experience	Most of the general public had never seen them in the wild except flying overhead and had little contact with geese	Participants in the Strathbeg and Islay groups were more knowledgeable about geese and their impacts.

Impact on their lives?	Most of the general public exhibited a general indifference towards the wild geese as they did not make any impact on their day-to-day lives. When views were expressed they differed from positive ("nice for them to exist", "hearing them is quite special") to fairly negative (e.g. "couldn't care less", "smelly creatures").

Discussions about "visitors" and "the economy" came up repeatedly and participants generally concluded that if the presence of geese do bring in tourists everyone benefits to some extent by the money that comes into the country. | Strathbeg and Islay resident groups had stronger views about geese and felt the *"presence"* of the wild geese added something indirectly to their life in the region. Some thought their impact was positive, others negative. Participants talked about everything from *"enjoying the geese and seeing a skein fly over"*, to putting them *"pretty far down on the list of things that added quality to their life"*

Strathbeg visitors said geese have connotations with changing seasons, nature, and farming: "lovely it is to see them arrive each year ... marks the passing of the seasons". |
| *Topic: Goose conservation as a government priority* | | |
| **How important?** | There was a range of opinion but many participants felt that the government had other more important issues to deal with. Some felt the goose issue appeared to be quite localised and best dealt with at local and regional levels | Islay residents had a more positive attitude towards government intervention on the geese issue than other groups but qualified this by saying an increase in the goose population will *"put things further out of balance than they already are."* |
| **What is the government view?** | Most people considered the main reasons for protecting the geese to be concerned with global rarity, local economic impact of shooters and birdwatchers, impacts on agriculture.
Also mentioned within the discussion that took place in many of the groups was the issue of *"pressure being applied by other governments"* (i.e. international and EU regulations). | |
| *Topic: Target goose populations* | | |
| **Target populations?** | Most groups felt that Pink-footed and Greylag populations could be reduced, but increases in the rarer species such as the Greenland Barnacle and White-fronted desirable. | Islay residents did not want populations reduced but were anxious that geese numbers did not increase.

Several participants in all groups felt unable to make up their mind, stating this was a job for *"the experts"* or that it was difficult to *"pluck a number out of thin air"*. |

17

Minimum population?	Most participants indicated a desire for all populations to be kept above the level that may threaten the geese or put them at a risk of extinction. Many participants across Scotland thought that it would have be helpful to know what is considered a *"safe minimum number"* for each species.	
Means of control	Most groups took issue with any control mechanism that caused direct harm to the geese.	Strathbeg residents suggested a cull on Pink-footed and Greylag species
	Topic: Farmers Right to Protect Crops	
Do they have the right?	Across focus groups most participants believed that farmers had a right to protect their crops *"because this was their livelihood"*	
How should they be controlled	Participants generally favoured *"scaring off"* the geese or *"distracting them"* from crops, as opposed to harming the geese in any direct way. Some thought that geese might be able to be controlled *"by making the bird reserves so attractive that the geese would not go seeking farmers fields"* or by changing agricultural practice.	A few participants in every group who felt that *"culling in specific areas may be beneficial"* or that a *"limited cull should be undertaken for certain species"*. This was most strongly expressed in the Strathbeg resident group.
	Topic: Compensation Payments	
Awareness?	Participants in the general public groups, were generally not aware that some farmers received government payments for damage to crops caused by geese.	Islay and Strathbeg resident groups were more aware.
Is it fair?	Across groups participants believed that *"it was fair for farmers to be compensated for crop damage"*, however many participants felt that if other control methods (see above) were implemented there may not be a need to give farmers compensation. Farmers could do more to help themselves. In some groups, discussion turned to the issue of *"over abundance"* of crops and feeling that the pressure on farmers to *"continually produce maximum yields and plant fields in winter has helped to create the current problem"*.	The Islay resident group were not sure that small farmers were getting a *"fair deal"* and thought that *"the bulk of payments were going to large farms and the smaller farmers were loosing out."* (Much time was spent by the Islay group discussing the inequities of the current system of payments!).

Should tax-payers money be used?	Considerable debate about the use of tax-payers money to fund compensation. Questions raised included: *"How is it policed?"* *"How much is it spent and to whom does it go?"* *"What are other countries doing?"* *"What about other compensation that goes to farmers?"* *"Are farmers getting double subsidies through crop damage compensation and other programmes such as set aside?"*	There was no support from members of the Islay resident group for the current policy of compensation coming from general tax funds. Nearly every participant mentioned RSPB as having some responsibility for assisting with crop damage compensation payments.

Topic: Willingness to pay

Are you willing to pay?	Common to nearly every participant across focus groups was the lack of "willingness to pay" (WTP) though additional taxation to support probable increases in compensation payments. This was partly due to the preceding conversation on alternatives to tax and as one individual stated: *"We pay enough taxes for all kinds of things; probably the money would be better spent in other areas like education and NHS"*	
How much are you WTP?	In response to the *"If yes, how much?"* question, the range considered was between .03 pence and £5.00 pounds.	
Why not WTP?	Some wanted more information before deciding, others said there were more important priorities. Many stated WTP for goose conservation was conditional on how it was paid for. One woman from a farming family stated *"…I'm a farmer's daughter, and I think that farmers today get so many forms of compensation, not like when I was growing up. I love nature and try to educate my children on the value of birds and other wild animals but I am not certain that I feel that the geese are so threatened or that we should be paying for compensation through extra taxes so I would have to say No as well."*	Strathbeg visitors said they were not WTP because enough geese already.

4.3.5   The aim of this study was to gain some further insight into the importance of wildlife and goose-related tourism in the local economies of Islay and around the Loch of Strathbeg particularly with respect to the influence of future conservation management on tourism. As quantitative estimates of the economic impact of geese-related tourism at either location was not required, it was decided to investigate this topic and related issues using Focus Groups.

4.3.6   In April two Focus Group meetings were held to discuss the economic importance of wild geese to the local economy. The Islay meeting was held at the Port Charlotte Hotel and the Strathbeg meeting at the Waterside Inn, on the outskirts of Peterhead. There were 6 participants in each group drawn from a range of tourist businesses including accommodation providers (e.g. hotels, bed and breakfast etc.), restaurants, pubs, shops and tourist attractions. Some of the participants (but less than half), were known to have direct links with wildlife tourism.

4.3.7   The meetings took the form of a semi-structured discussion led by a moderator. The discussions focused on general tourism in the area, the importance of wildlife tourism, and the level of tourism activity associated with wild geese in particular. The main findings of the discussions are summarised in Table 4.2.

4.3.7   Key findings were:

- Tourism is relatively more important to the local economy of Islay than to the Loch of Strathbeg area.
- Geese play an important role in the tourist economy of both places: Goose-watching is of some importance to bird-watchers on Islay, whereas goose shooting is more important around the Loch of Strathbeg.
- Moderate fluctuations in the size of the goose population were not thought likely to affect tourism in either area.
- Commercial shooting is not considered to be an option at the moment on Islay, but there was widespread agreement that the local economy of Strathbeg would benefit from extending the open season to February.

Table 4.2 Summary of Tourism Group Discussions

Topic: General Tourism

Questions…	Islay	Strathbeg
General image?	All participants considered Islay to be a special place because it provided a unique combination of wildlife, friendly people, attractive countryside, and of course whisky.	All participants felt that the area was a forgotten corner of Scotland, bypassed by tourists from the south.
Who comes?	Family holidays are popular (*families who come here say their children are safe and that's a big plus in this day and age*). There are also a lot of 'enthusiasts' who come for the wildlife and for the whisky.	Visitors often had local connections and/or interested in tracing their ancestors ('the graveyard trail'). Other activities included coastal walking on the beaches, visiting castles, and enjoying the peace and quiet.
Seasonality?	Everyone agreed that tourism was highly seasonal. The tourist season had gradually extended over the last 20 years, mainly because of wildlife visitors.	Everyone agreed that tourism was seasonal but there was a strong winter period when goose and wildfowl shooters arrived. The poorest months were February and March.
Main limitations?	The main restrictions on developing tourism include the high costs of fuel and public transport. There are few facilities for back-packers. All participants were frustrated with the lack of co-ordinated marketing.	Main problems facing tourism were high costs of transport and poor marketing of the area. *'No one knows about us up here ….. even in Aberdeen you cannot find information for visitors to this corner'*. Another problem was the absence of big attractions or nice landscapes: *'We had a big cruise ship that docked at Peterhead but they just got on the bus and drove to Deeside'*.

Topic: Wildlife and Tourism

Questions…	Islay	Strathbeg
Who is attracted by wildlife?	Islay is very popular with wildlife-watchers and sportsmen. The latter are less well integrated in local economy. As one participant declared '*the (sporting) estates tend to look after their own people*'	Wildlife is important to the area, particularly the goose shooters in the winter period.
What about income from shooters?	Some participants said that significant income in the winter months had been lost when let goose shooting was banned.	Shooters tend to spend a lot more money in the local economy than other visitors, staying in the best hotels and purchasing expensive gifts.

21

Why do birdwatchers come?	Most visitors are attracted by the diversity and abundance of wildlife rather than particular species. Two different types of birdwatcher: *the 'twitcher' who ticks them off on a list and doesn't want to stop and look at them ......and there's the other kind who will stop and watch the ducks or heron or whatever is on the shore* People who come in the summer often like to do a bit of birdwatching but they also do other things such as walking. Specialist bird-watchers come in winter to the reserve where they can see a range of birdlife such as swans, eider duck and other wildfowl. Visitor numbers increase when there is a rare migrant such as the flamingo that was seen several years ago.
What about income from birdwatchers?	The stories about birdwatchers bringing their own food and sleeping in their car was viewed as being a stereotype and outdated: wildlife visitors spend money just like anyone else. Birdwatchers are much more likely than other visitors to camp or caravan, and to bring their own food. As a result they have a bit of a reputation for not spending much money locally.

Topic: Wild Geese and Tourism

Importance of geese to the local economy	Most agreed that people did come just to see the geese, especially in October when the geese arrive and in April/May when the geese depart. These were 'specialists' with *'a certain cultural and academic inclination .... and a distinctive age range (30-50)'*. Geese-watchers often come in organised parties and there are a number of small bus/tourist companies that bring visitors to see the geese. Unfortunately these companies are not based on the island. Wild geese were the main quarry species of sportsmen and hence are very important. However, the group did not think that many birdwatchers visited specifically to watch geese. Geese were one of many attractions at the reserve - swans were a bigger attraction. (One person thought that the only person who specifically watched geese was the RSPB warden!)
Impact of population changes on local economy?	None of the participants thought that any plausible or foreseeable changes in the wild goose population would make much difference to the number of people visiting the island for goose watching. Everyone agreed that even substantial changes in the goose population was unlikely to affect the number of birdwatchers coming to the area. The impact on shooting is more complex but a local goose guide felt changes in the goose population would make little difference to the income from shooting because the main limitations on shooting are access to gun sites and the problem of disturbance.

Changes in the control of shooting?	Several people expressed concern that the re-introduction of commercial shooting would generate bad publicity and could threaten the islands reputation among birdwatchers: *'the trouble is the media will get hold of it and blow it up into a huge thing. Someone will come along, another Bellamy perhaps, and say they're shooting the geese again, oh dear, oh dear'* Granting farmer shooting licences to control damage by geese was different and most agreed that this policy if carefully explained would have no negative impact on tourism	Many people felt that the best way to increase the economic benefits of shooting would be to alter the open season from September-January to October-February. September is not a good month for shooting and there are still a number of summer tourists around at this time and it is the worst month for tourist businesses. In February the geese are still around and it is the worst month for tourist businesses. Extending the season to March would also help reduce damage to agricultural crops. There was disagreement over the question of legalising the sale of meat with some arguing that it would boost the local hotel and shop trade. Some believed that poaching would increase and this would lead to a reduction in the goose population and adversely affect let shooting.
Do farmers benefit from tourism?	Although some farmers ran B&B enterprises, there was a view that farmers did not benefit from tourism in general. A farmer in the group mentioned a shortage of capital to invest as one of the main barriers to further involvement.	Some farmers benefit from the geese directly from leasing shooting rights. However, they are not very active in the tourism industry, preferring to concentrate on growing crops.

CHAPTER FIVE WILLINGNESS TO PAY

Summary

This chapter describes the main results of the willingness to pay surveys to estimate the non-market benefits of migratory wild geese.

The Chapter describes the main results for the two main valuation techniques employed.

1. Contingent Valuation (CV): which provides estimates of WTP for specific management plans.
2. Choice experiments (CE): which provides estimates of WTP for specific attributes of goose management policy.

The results of a validation exercise for these estimates are also described, including results from the Market Stall (MS) experiment.

A detailed report of the study can be found in Technical Report B.

Key Findings

General Attitudes

- Geese conservation did not seem very important relative to other nature conservation issues, being ranked second last of six, with only "re-introducing beavers" coming lower. This was the case for all groups surveyed.

- One-third of residents felt that the advantages of wild geese feeding in the area outweighed the disadvantages, with 14% taking the opposing view.

- Only 8% of visitors said that the opportunity to watch geese was 'very important' in terms of their reasons for visiting the area

CV results for general public
(all values in £ per household per year for 10 years)

- Policy options investigated were:
  A: Goose numbers to be controlled by habitat management only (no shooting)
  B: Management to avoid a 10% fall in endangered species
  C: Management to obtain a 10% increase in endangered species only
  D: Management to obtain a 10% increase in *all* wild geese species

- More than 50% of people supported each of the four options to protect or enhance wild geese populations.

- Mean WTP per household was as follows:

  Option A: £8   Option B: £11   Option C: £16   Option D: £21

CE results for general public, local residents and visitors
(all values in £ per household per year)

- General public: the only significant policy attribute was whether geese are controlled by shooting or not. On average households would be WTP almost £10 more for a policy that does not involve shooting.

- Islay residents: Whether geese are controlled by shooting or some other means is unimportant. However, they would pay almost £30 to avoid a big increase in goose numbers (+50%) and pay an extra £12 for policy that protected endangered species only (rather than all species)

- Islay visitors: Among this group there was support for a policy that focussed on endangered species only (+£16). They also prefer policies which avoid shooting as a means of control (+£7), and which target conservation at all sites in Scotland (+£7), rather than in special areas only. Finally, whilst they have a significant and positive preference for a 25% increase in geese numbers (+£15), they have a negative preference for a 50% rise. Thus, even this most pro-geese group are against large increases in goose numbers.

- Strathbeg Residents: None of the attributes except for tax cost were significant.

- Strathbeg Visitors: Preferences are rather similar to Islay visitors although less marked. They thus significantly value a policy which avoids shooting as a means of control (+£19), and which focuses on endangered species (+£8). The stronger preference for a policy that avoids shooting may reflect unease about the intensity of sport shooting that currently takes place at Strathbeg.

Validation of willingness to pay surveys

- Individual WTP responses were consistent with prior expectations (that is, made sense from an economic point of view!), with WTP positively correlated with income; how important the respondent thinks protecting wildlife is relative to other goals of rural policy; whether the respondent has seen wild geese; and which policy scenario was being valued.

- A comparison of mean WTP estimates, derived from the CV and CE surveys, for a policy to stop shooting was made to test whether the two techniques generate similar benefit values. The CE estimate (£9.23) was found to lie within the 95% confidence interval of the CV mean estimate (£8.32). This is a re-assuring finding.

- Feedback sessions were organised to allow people who participated in the CV and CE surveys to assess the reliability of the exercise. While most people could recall details of the policy options and were confident that they understood the valuation question, others complained of inadequate information and time to make a decision.

Market Stall (MS) Experiment

- The MS approach provided participants with more time and information to come to a decision about their WTP than is possible in surveys. Compared to the main CV survey, the MS approach generated significantly lower estimates of WTP, with survey estimates on average 3.5 times greater than those obtained from the MS.

- There was evidence that the MS estimates are more reliable than the CV survey estimates. For example, protest response rates were lower and more of the variation in individual WTP could be explained by socio-economic factors we would expect to influence WTP.

Aggregate benefit estimates

- Below we estimate the following annual, aggregate benefits (based on a figure of 2,186,500 households in Scotland) and the CV survey sample estimate of WTP calibrated downwards by a factor of 3.5 (based on the MS result):
  Project A: £5.2 million to prevent the shooting of geese
  Project B: £6.8 million to prevent a 10% fall in endangered species
  Project C: £10.2 million to achieve a 10% rise in endangered species
  Project D: £13.1 million to achieve a 10% rise in all geese species.

5.1   The Nature of Benefits

5.1.1   Policies for the conservation of wildlife produce many types of benefits. For example, additional tourist spending in rural areas. One important class of benefits, and the one we focus on here, are described by economists as "non-market". This means that even though people benefit from geese conservation, these benefits do not show up in market prices or in financial returns. For instance, people who are concerned about the fate of wild geese would benefit (in terms of "increased satisfaction") if they felt a policy would increase the chances of conserving geese, even if these individuals never travel to Islay to go bird watching. Economists use the concept of Willingness to Pay (WTP) to derive a monetary measure of what this increased satisfaction (policy benefit) is worth to people. Monetary measures of these non-market benefits can then be compared with the monetary costs of the scheme to see if it is worthwhile on cost-benefit grounds.

5.1.2   Much experience now exists in the UK with estimating WTP for non-market benefits, in contexts as varied as low flow alleviation in rivers and landscape improvements in forests. Two main types of method are available: those relying on peoples' *stated preferences* for environmental changes; and those relying on *revealed preferences*. In this project stated preference methods are used since these are capable of capturing the widest range of beneficiaries from conservation policy. Within the field of stated preference methods, two main options are those used here: contingent valuation (CV) and choice experiments (CE).

5.1.3   CV relies on surveys to find out how most people would be WTP for changes in the condition or management of environmental assets through directly asking them. In our application, respondents are asked the most they would be WTP for different changes in, for example, the number of endangered geese conserved. This yields estimates of the non-market value of this change. The word "contingent" is used since these WTP estimates are derived contingent on a description of what the planned change is, and how people can pay to either have it go ahead or to avoid it. CV has also been widely used in the UK, especially in producing estimates of the economic value of nature conservation. For a simple description of the method, see Hanley and Spash, 1993. (For a detailed review, see DETR, 2001).

5.1.4   CE are a development of an approach which has been used for some time in transportation and marketing research. It consists of asking people to make choices between alternative descriptions of environmental goods or policies. These descriptions are set out in terms of the *attributes (characteristics)* of the environmental good or policy, and the levels that these take. For instance, alternative forest designs could be described using the attributes of species diversity, age diversity and recreational facilities. Different levels could be ascribed to these attributes, such as setting species diversity at levels of 100% conifer, mixed conifer and broadleaves, and 100% broadleaves. By adding an attribute which shows the cost to the individual of each alternative option the economic value of a change in any of the attributes can be worked out; in the case above, this could be the implied taxpayer cost of different forest designs. For an introduction to the use of choice experiments in environmental valuation, see Hanley, Mourato and Wright (2001).

5.2   Contingent Valuation Results

5.2.1   The CV study was used to estimate the benefits of four different policy options:

   A: Goose numbers to be controlled by habitat management only (no shooting)
   B: Management to prevent a 10% fall in endangered species
   C: Management to obtain a 10% increase in endangered species only
   D: Management to obtain a 10% increase in *all* wild geese species.

5.2.2   In total, 419 responses were obtained, all from members of the general public in Scotland. People were first asked about their attitudes to nature conservation as part of rural policy, and their attitude to wild geese compared with other conservation issues. Main findings were as follows:

   •   Protecting wildlife was ranked 3rd amongst the many objectives of rural policy, lower only than providing employment and producing food (Figure 5.1)
   •   Goose conservation did not seem very important relative to other nature conservation issues, being ranked second last, with only "re-introducing beavers" coming lower (Figure 5.2)
   •   Farmers being paid to manage wildlife was supported by the majority. However only a minority agreed with the view that farmers should be compensated for damage due to wildlife, or should be allowed to kill wild animals (Figure 5.3).

Figure 5.1: Percentage of people who gave top-ranking to each of the listed countryside policy priorities (general public)

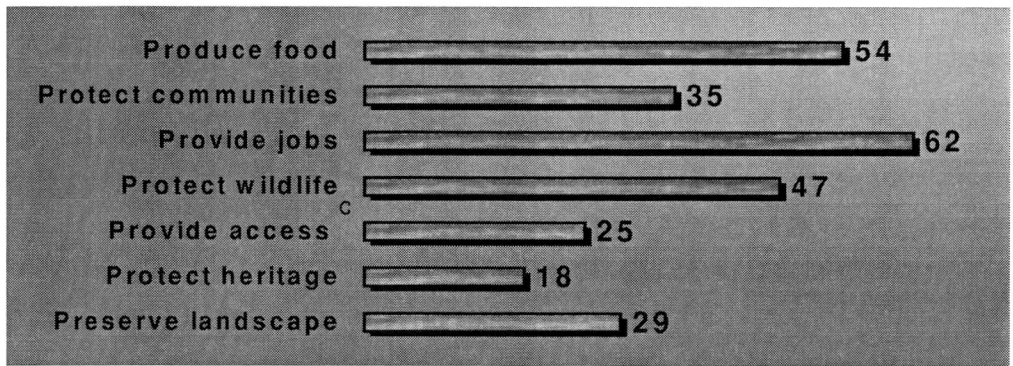

Figure 5.2 Percentage of respondents who gave top-ranking to each of the listed wildlife policy priorities (general public)

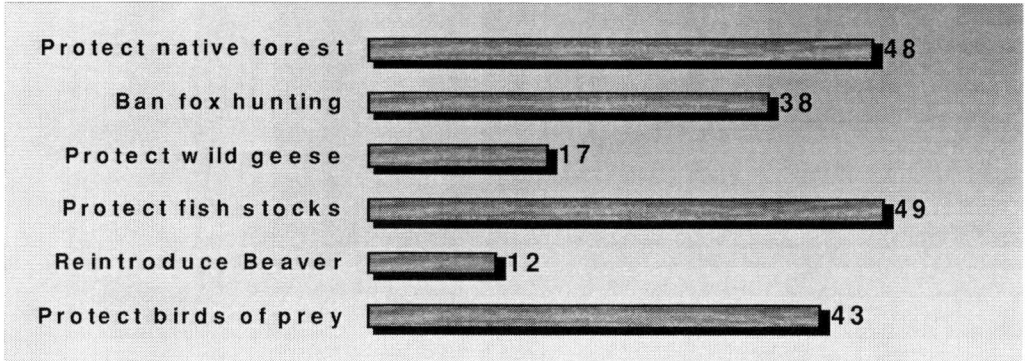

Figure 5.3   Percentage of people who strongly agreed with the following opinions on farming/wildlife interactions (general public)

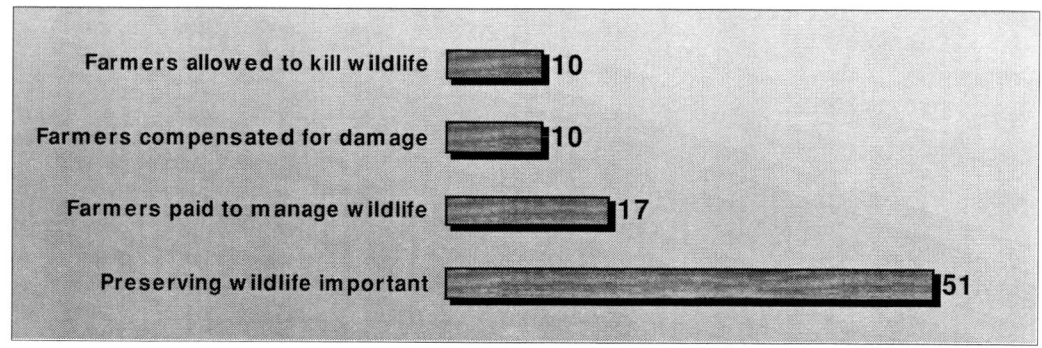

5.2.3 A rather small fraction of our sample had ever visited Islay or the Strathbeg areas. However, most people said they were willing to support policies for geese conservation, even if these were costly. The level of support across policy options varied, as may be seen in Table 5.1.

<table>
<tr><th colspan="5">Table 5.1 Level of support for different geese conservation policies
(% in each category)</th></tr>
<tr><th></th><th>Option A</th><th>Option B</th><th>Option C</th><th>Option D</th></tr>
<tr><td></td><td>No shooting</td><td>10% fall in endangered</td><td>10% rise in endangered</td><td>10% rise all species</td></tr>
<tr><td>Yes</td><td>66</td><td>59</td><td>51</td><td>53</td></tr>
<tr><td>No</td><td>25</td><td>20</td><td>35</td><td>34</td></tr>
<tr><td>Unsure</td><td>9</td><td>21</td><td>14</td><td>13</td></tr>
</table>

5.2.3 However, all these figures tell is the *direction of preferences:* whether people support a policy or not. For economic analysis, we need to know *by how much* people support the policy options, in other words, we need to know their WTP for each option. This was investigated by asking each respondent eight "payment" questions. In each of these, people were presented with a possible cost to them of the policy, in £ per household per year over a 10-year period. Tax payments were used since such policies are paid from general tax revenues

5.2.4 Five responses to each payment level were possible: Definitely Would Pay (DWP); Probably Would Pay (PWP); Not Sure (NS); Probably Would Not Pay (PWNP); and Definitely Would Not Pay (DWNP). The payment question differed depending on the project scenario.

5.2.5 For example, for the "no shooting" case, people were told:

> *"At the moment the government is considering the future policy for managing wild geese by no longer allowing any species of geese to be shot. Instead, all species will be controlled through habitat management. It is not expected that this change in policy will affect the total number of geese. However, it would cost the tax-payer more money because habitat management is a more expensive option than shooting and will likely result in greater damage to agricultural crops."*

They were then told:

> *"Imagine that the additional costs of the project resulted in your household having to pay more tax each year for the next ten years. I would like you to think about how much your household would be prepared to pay. To help I will read out some possible increases in taxation for this new management policy. When considering how much you would be willing to pay, remember that the revenue from the tax payment would **only** be spent on goose management, what you can afford, and that there are other things that tax money could be spent on"*

5.2.6   Finally, the eight possible payment amounts were read out in random order, and the respondent's answers recorded. Using these responses, we can estimate mean WTP for each policy option. In order to be conservative, those amounts which respondents said they would definitely (rather than probably) be willing to pay are used.

5.2.7   In Table 5.2, mean WTP and the 95% confidence interval, together with the median WTP amount are rounded to the nearest £ for each option. The mean amount is what, on average households would be willing to pay. The median WTP can be thought of as the highest amount that at least half of the people in our sample would vote "yes" to. As may be seen, the highest value measured in this way is attributed to a policy to increase all geese species by 10% (Option D). However, as WTP for this option is not significantly higher than the WTP estimate for Option D, which concerns endangered species only, we can conclude that there is no evidence in a statistical sense to suggest that people value non-endangered species.

Table 5.2 Willingness to Pay estimates from the contingent valuation survey (All values in £/household/annum)			
	Mean WTP	95% Confidence Interval	Median WTP
Option A: no shooting	8	5.58-11.06	3
Option B: Prevent 10% decline in endangered species	11	6.34-15.63	4
Option C: Obtain 10% increase in endangered species	16	6.30-26.26	5
Option D: Obtain 10% increase in all geese	21	10.96-30.98	9

5.3   The Choice Experiment Results

5.3.1   In any choice experiment, the most important tasks are to select the attributes which will be used to describe the environmental asset (in this case, wild geese). Following focus group discussions (see Chapter 4) and pilot survey work, the attributes and levels shown in Table 5.3 were chosen.

Table 5.3 Attributes and levels used in the Choice Experiment	
Attribute	Level
Species	Endangered species only
	All species
Location	Special reserves only
	All locations in Scotland
Method of Control	Habitat management only
	Shooting and habitat management
Population Change	Small fall (-10%)
	Stay the same (0%)
	Small rise (+10%)
	Moderate rise (+25%)
	Large rise (+50%)
Tax	£1; £5; £10; £20; £35; £60

5.3.2. This mix of attributes is intended to capture those features of the goose management "problem" that government is able, at least partially, to influence through policy design, as well as the costs of policy to the taxpayer. The next task was then to combine these attributes and levels into a series of choice pairs. An example is given in Figure 5.4.

Figure 5.4: An Example Choice Experiment

Please consider the following options:

Policy A	**Policy B**
Species protected by policy:	Species protected by policy:
Endangered species only	**Endangered species only**
Means of control:	Means of control:
Habitat management & Shooting	**Habitat management & Shooting**
Location:	Location:
Special reserves only	**All sites in Scotland**
Population change over 10 years:	Population change over 10 years:
Stay the same	**Moderate rise (25%)**
Price per year to you over the next ten years in extra taxes: **£10**	Price per year to you over the next ten years in extra taxes: **£60**

Response: I would choose Policy A   _____
              I would choose Policy B   _____
              I would choose neither Policy A nor Policy B   _____

5.3.3   Respondents were asked to complete four of these choice tasks. The basic idea is that by making such choices, people reveal three things: (1) the relative values they place on the different attributes (e.g. where geese are conserved versus how they are controlled); (2) which attributes they care about significantly in a statistical sense; and (3) the money value of changes in any of the attributes (for example, WTP for a policy which resulted in a 10% increase in geese relative to the current situation). These WTP amounts are estimates of underlying utility changes for the people concerned: something which makes them feel better off is associated with an increase in utility; something that makes them feel worse off, with a decrease in utility. Estimates of these three types of insight can be got from running a special type of regression model which relates people's choices to the attributes and the levels they take.

5.3.4   First, though, let us look at who was sampled in the choice experiment, and at their general attitudes and behaviour. Five populations were targetted:

- the general public (426 people sampled)
- residents of Islay (205)
- residents in Strathbeg (196)
- visitors to Islay (212)
- visitors to Strathbeg (202)

5.3.5   In total, some 1,241 people were interviewed. With regard to attitudes to environmental issue, we focus on the visitors and residents samples, as the general public sample was rather similar in views to that of the CV survey. Residents of Islay and Strathbeg rate wildlife conservation as only 4th in importance compared with the general public (CV sample) who rated it 3rd, pushing "protecting rural communities" higher (Figure 5.5). Wild geese conservation was again rated as of low importance compared with other nature conservation objectives, coming next to last. Compared with the general public, local residents were more in agreement with the views that farmers should be compensated for damage caused by wildlife (58% agreement) and that farmers should be allowed to kill wild animals (49% agreement), as Figure 5.6 shows. Finally, one-third of residents felt that the advantages of having wild geese in the locality outweighed the disadvantages, with 14% taking the opposing view.

5.3.6   The views of visitors to Islay and Strathbeg were somewhat different to those of residents, as might be expected. Visitors gave a higher ranking to countryside management as a goal of government policy, and placed a higher importance on protecting wildlife as a goal of countryside policy, and a lower importance on providing employment and producing food. Within wildlife policy, visitors had an almost identical ranking to residents. Protecting fish stocks and native forests are ranked top; geese conservation was ranked 5th. In relation to farming/wildlife interactions, visitors were less willing than residents to support farmers' rights to kill wild animals causing damage, and gave more support for wildlife protection (Figure 5.7).

Figure 5.5: Percentage of people who gave top-ranking to each of the listed countryside policy priorities (visitors and residents)

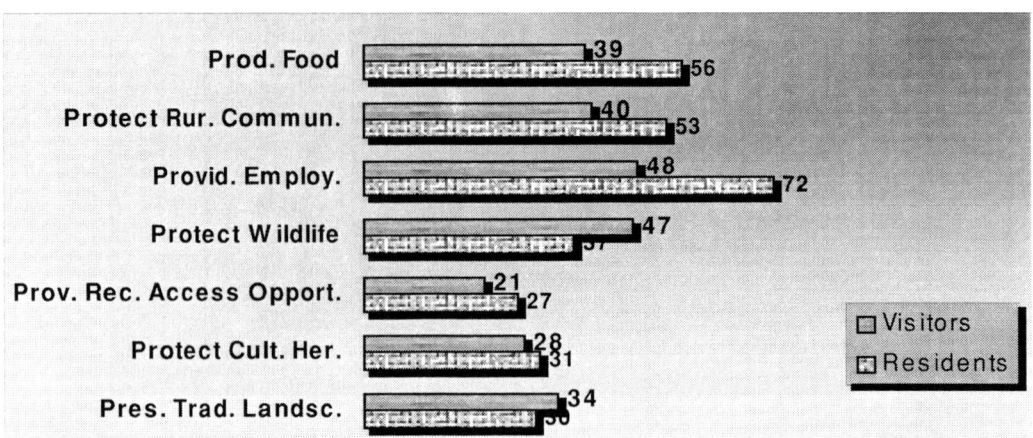

Figure 5.6 Percentage of respondents who gave top-ranking to each of the listed wildlife policy priorities (visitors and residents)

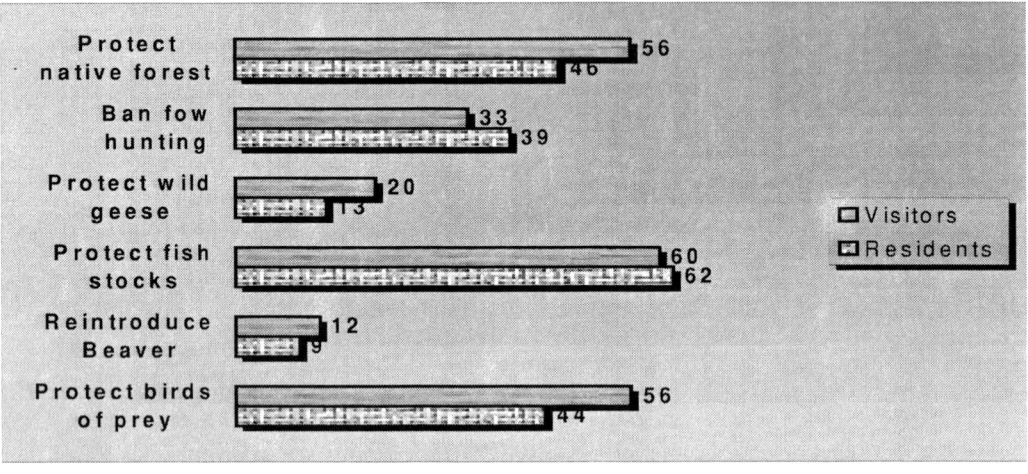

Figure 5.7   Percentage of people who strongly agreed with the following opinions about farming/wildlife interactions (visitors and residents)

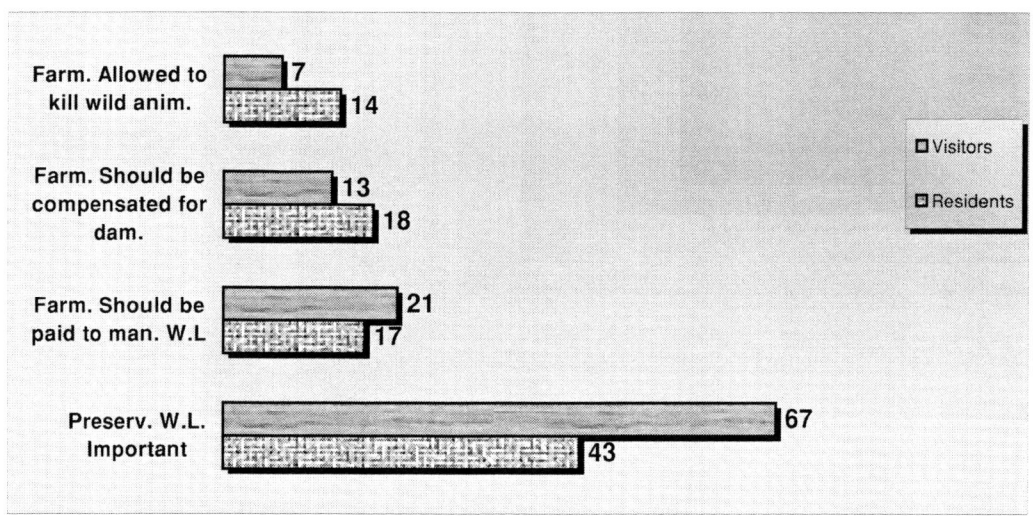

5.3.7   Visitors were also asked a series of questions about their reasons for visiting the area, the number of times they visited and their place of origin. A big difference was apparent between Islay and Strathbeg in terms of where visitors came from. On Islay, 91% were non-residents staying at least one night: this figure fell to 27% for Strathbeg. Some 27% of visitors to Strathbeg were locals (3% on Islay), whilst 40% were non-local day-trippers (6% on Islay). Finally, only 8% of visitors to both locations said that the opportunity to watch geese was "very important" in terms of their reason for visiting the area. For over half (56%), it was "not important at all".

5.3.8   As stated above, data from the choices people make in a CE can be used to look at which attributes are significant to them, their ranking of these attributes, and their WTP for changes in attributes. An important first step was to test whether the five samples were sufficiently similar in their responses to the choice questions for the data to be pooled. The result obtained was that the samples are statistically all different from each other in terms of their preferences for geese conservation policy: the general public has different preferences to residents or visitors to the two areas, and residents are different to visitors. Also, Strathbeg visitors are different to Islay visitors, a conclusion which also holds for residents. We thus had to estimate different models for each of the five groups in the overall sample, relating their choices to the attributes and the levels they take. It should be emphasised that what follows is based on the numerical results from estimating these models.

5.3.9   Which attributes were relevant to peoples' choices? In summary, the main results were:

- for the general public, the only significant attribute was whether geese are controlled by shooting or not. People's utility goes down when geese are shot. This means that *where* geese are conserved and *which* geese are conserved does not seem to be important to the general public.

- for Islay residents, a different picture emerges. Whether geese are controlled by shooting or some other means is unimportant. Big increase (+50%) in geese numbers are viewed negatively, but reductions in geese numbers below the current level would be regarded as a loss. Finally, Islay residents would rather protection focussed on endangered species rather than all species.

- for visitors to Islay, a significant desire for conservation to be focussed on endangered species was also found. They also significantly prefer policies which avoid shooting as a means of control, and which target conservation at all sites in Scotland, rather than in special areas only (logical, since they probably do not live in these special areas!). Finally, whilst they have a significant and positive preference for a 25% increase in geese numbers, they have a negative preference for a 50% increase. Thus, even this most pro-geese group are against large increases in the goose population.

- For residents in Strathbeg, none of the attributes except for the cost of the policy were significant.

- For Strathbeg visitors, preferences are rather similar to Islay visitors although less marked. They thus prefer a policy which avoids shooting as a means of control, and which focuses on endangered species. (N.B As interviews were carried out in May and October, this represents the views of non-shooters). A reduction in current goose numbers would significantly and negatively affect Strathbeg visitors, but there is no significant support for a policy to increase geese numbers.

5.3.10 In Table 5.4, the impacts noted in the above bullet points are converted into WTP amounts, but only where a statistically-significant effect exists. This is accomplished by using the relationship between the cost of the policy and support for policy uncovered by the choice experiment. How this was done is described in more detail in Technical Report B.

Table 5.4: WTP estimates from choice experiment by population sampled	
Sample	*Willingness to Pay estimates (£ per household per year)*
General Public	Stop shooting: £9.23
Islay Residents	Target endangered species only: 12.26 Avoid 10% fall in population: 24.98 Avoid increase in population by 50%: 29.67
Islay Visitors	Stop shooting: 6.74 Target endangered species only: 16.50 Target all sites rather than just special reserves: 6.73 Obtain increase population by 25%: 15.39
Strathbeg Residents	No significant effects
Strathbeg Visitors	Stop shooting: 19.28 Target endangered species only: 7.72
All visitors	Target endangered species only: 12.13 Target all sites rather than just special reserves: 6.03 Stop shooting: 12.24
All residents	Target endangered species only: 7.55 Avoid 10% fall in population: 13.86 Avoid increase in population by 50%: 13.32

5.4 Validation

5.4.1 How can we tell how valid these estimates of economic benefits are? One way would be to compare results for WTP for wild geese conservation from this study with those from other studies. Unfortunately, this cannot be done since this is the first UK study of this type. Another option is to compare the way in which the surveys were carried out relative to best practice guidelines. As may be found by referring to DETR (2001), the CV and CE studies do indeed adhere to these guidelines.

5.4.2 Several other forms of validation are possible. First, we can look at the internal validity of the WTP responses. The usual way of doing this is to statistically relate these responses to variables thought likely to influence WTP, and see whether the statistical relationships are in accord with a prior expectations (that is, make sense!). Full details may be found in the technical report. Summarising, however, WTP bids for geese conservation depends significantly in statistical terms on the following variables:

- income: richer people are willing to pay more for conservation
- how important the respondent thinks protecting wildlife is, relative to other goals of rural policy
- whether the respondent has seen wild geese
- which policy scenario the respondent was being questioned about.

5.4.3 A second option is to compare WTP estimates from the CE and CV surveys. As the CV study was done on members of the general public only, we restrict attention here to CE results from this sample. The only comparison which it is strictly correct to make is with estimates for WTP for a policy to stop shooting, as this was the only relevant attribute in the CE general public sample. Comparing the two sets of results shows the following:

- CV mean WTP: £8.32/household/year.
- CE mean WTP: £9.23/ household/year.

5.4.3   The CE estimate lies within the 95% confidence interval of the CV mean estimate. This implies that the two estimates are not different statistically at the 95% level of confidence. This is a re-assuring finding.

5.4.4   A third form of validation involved verbal feedback from a small sample of survey respondents. These respondents were asked to discuss critically the valuation exercise they had completed and to assess the reliability and accuracy of their responses. They also provided a forum for the format and content of the survey to be discussed and generated information that made interpretation of the WTP data easier.

5.4.5   System 3 attempted to recruit participants for four feedback sessions. A 'local resident' and a 'visitor' group were convened in May 2000, and recruitment for two general public feedback sessions took place in August 2000. Unfortunately, it proved impossible to recruit enough people for the general public due to the geographic spread of the sample, hence feedback took the form of telephone interviews.

5.4.6   A summary of the findings from the feedback research is presented in Table 5.5. The main conclusions were as follows:

- Most respondents thought it was important that the views of the public are canvassed regarding conservation issues and supported the idea of WTP surveys in the context of government policy for conservation and wildlife management. (Although the role of experts in policy development was not disputed).

- Although respondents recalled details about the policy option they had been asked to consider, the details they remembered were often incorrect. This raises some doubts about the validity of the survey responses.

- A key finding was that people with a working knowledge of the topic in hand wanted to be presented with more information before making their willingness to pay decisions. Such people were more likely to question the detail of the policy options whereas people with little or no previous knowledge about wild geese appeared to take the willingness to pay choices at face value.

- Several respondents expressed disquiet about the time they had to digest the information and to discuss it. For example, one woman was interviewed just after returning home from dinner with friends, and did not concentrate because her friends were waiting on her. Another said she would have been more confident about her responses had she had the chance to think and ponder: she explained that she had answered in the survey that she had never seen wild geese before but later realised she had seen them flying overhead.

- A more positive conclusion for hypothetical valuation approaches was that tax level was an important factor in the decision of almost all participants.

5.4.7   The fourth approach to the validation of the WTP results was the implementation of a Market Stall experiment. The methodology and results from this exercise are discussed in detail in the next section.

5.5   The 'Market Stall' experiment

5.5.1   Earlier the results of WTP surveys using the CV and the CE approaches were described.   Both studies were implemented using in-person interviews carried out by professionally trained staff as this is the recommended approach (NOAA, 1993). Typically, each interview lasted about 20-30 minutes, with interviews taking place in the family home or 'on site' (in the case of Islay visitors the latter included the ferry).

5.5.2   During the interview the respondent must assimilate information about the goose issue, search their memory for other pertinent information, integrate this into a judgement about their WTP based on their preferences and income, and communicate this judgement to the interviewer (Hanneman, 1994). There was evidence from the Feed-back sessions that this was quite a difficult task in relation to the conservation of wild geese.

Table 5.5: Summary of Feed-Back Focus Group Discussions

Question….	Converging Views	Diverging Views
Recall the questionnaire?	Although each person recalled slightly different rationales being behind the survey they had completed, they all recalled that it had been about wildlife and that the focus on wild geese had been clear.	
Recall choices? (CE only)	Most remembered that they had been asked to make a number of goose management policy selections, but none could remember the detailed contents of any of the choice sets they had been presented with.	
Was the task difficult? (CE only)	When questioned, no CE participant indicated that they had experienced conceptual difficulties in responding to the choice sets. However, when asked to complete the exercise again a difficulty emerged. Participants appeared to want to 'pick and mix' their own policy combination rather than deal with the choice sets in the requested manner.	
Were decisions dominated by one attribute?	Discussion suggested that no one attribute dominated policy choice in the CE. For example, the shooting option, although unpopular, had not led to the rejection of particular policy options. CV participants did not focus on any one aspect of the policy scenario presented to them.	On Islay, 'locals' more specific about the attributes that most influenced their choice but no common influence was identified. For some it was shooting, for others locational policy or species.
What about tax?	Most people had considered the tax level when making their choices. For example, a Strathbeg group member felt that she was already paying a considerable amount in tax and that the higher monetary values had put her off certain choices.	There was evidence that respondents were not clear as to the nature of the tax payment (personal contribution or what everyone should pay) and some concern that the additional tax might not go to the cause it was intended for.
Differences between Islay and Strathbeg	Local circumstances undoubtedly influenced the choices recorded by these two groups. For example, different attitudes towards shooting, bird reserves, and compensation payment mechanisms were apparent between residents of Islay and Strathbeg.	
More information needed?	In the Islay and Strathbeg groups knew what habitat management could involve, but wanted more details about what was on offer regarding habitat management in the policy options.	Members of the general public did not want more information: not very interested in the subject of geese.

5.5.3   The Market Stall (MS) is an alternative approach to estimating WTP that involves participants in two informal group meetings held approximately a week apart. Hence, compared to a survey, it provides a completely different decision-making environment. The main potential advantages of the MS are

- it provides participants with more time and information to determine their WTP
- gives participants the opportunity to discuss their WTP decision with the moderator and other group members
- provides the opportunity, during the week-long interval between the two meetings, for participants to re-evaluate their WTP following further thought, information searching, and crucially for household economic decisions, discussions with family members and friends

5.5.4   Due to the relatively small sample sizes involved in the MS experiment, only two out of the four CV management options were compared:

- **Option C:** which describes a plan to enhance the population of endangered species by 10% over the next 10 years through improved habitat management and stricter controls on shooting

- **Option D** which was identical to C, except that the populations of all four species of wild goose were to be increased by 10%.

5.5.5   As in the main survey interviewees were asked to indicate the degree of certainty they placed on being prepared to pay 8 different tax payments. Five responses to each payment level were possible: Definitely Would Pay (DWP); Probably Would Pay (PWP); Not Sure (NS); Probably Would Not Pay (PWNP); and Definitely Would Not Pay (DWNP). To avoid payment card effects only payment Card D was employed in the MS.

5.5.6   Meeting 1 was primarily concerned with the presentation of relevant information about the proposed project, described in an 'Information Folder', and a detailed explanation of the contingent market and means of payment. The information given in the folder was carefully designed to be understandable but was more detailed than would be possible in a survey context. Participants were given the opportunity to discuss any aspect of the project and to question the moderator. A 'Question and Answer Sheet' at the back of the folder was also provided to help clarify issues.

5.5.7   Meeting 1 (MS1) concluded with WTP being elicited using the same question format described for the survey, with one procedural difference: respondents were asked to write down their WTP and place their answers in a sealed envelope rather than respond verbally. This was done to provide respondents with some feeling of confidentiality. In addition, respondents were are also asked to complete the same set of basic questions about socio-economic status and attitudes to various environmental issues that appeared in the in-person CV survey.

5.5.8   During the week-long interval between MS 1 and 2 participants had the opportunity, if they so wished, to re-read the Information Folder, supplement their knowledge of geese, discuss the issue with relatives and friends, and crucially, to re-evaluate their WTP. In order to record their thoughts and activities related to the valuation issue participants were asked to complete a daily diary.

5.5.9   At Meeting 2 (MS2) participants were given the opportunity to ask further questions and to discuss any unresolved issues concerning the project. After the WTP question was repeated a de-briefing exercise was carried out to establish what influenced their decision.

5.5.10   A total of 52 people at four different locations were recruited for eight MS groups by the same market research company based on quota sampling approach used for the survey, with 43 returning for the second meeting. Of those who did not return, all except one individual who had left the area on business, completed the diary and second payment card over the phone, or by returning it in the mail. Statistical analysis of socio-economic data (income, household size etc) revealed no significant differences between the sample populations in the MS groups and the survey. The location and size of each group meeting is given in Table 5.6.

Table 5.6: Location of MS meetings and number of participants		
Location	Number: Session 1	Number: Session 2
Aberdeen 1	7	7
Aberdeen 2	6	6
Dumfries 1	7	5
Dumfries 2	6	4
Nairn 1	6	6
Nairn 2	8	7
East Kilbride 1	7	4
East Kilbride 2	5	4
Total	**52**	**43**

5.5.11   Analysis of diary entries for the intervening period between the two meetings revealed that participants had engaged in a wide range of actions that were of potential relevance to the issue of goose conservation and their WTP for the project. These included watching TV programmes, visiting local bird reserves, reading books and newspapers, and for most people discussing the project with colleagues, friends and family. A number of respondents noted questions they had about the management option such as uncertainty over project effects, and about the total cost of the project. A small number of people did not make any entries into their diary either because they were not really interested in the issue and/or had already made up their mind.

5.5.12   Figure 5.8 describes mean WTP for MS 1, MS 2 and the survey for the *'definitely would pay'* category[1] averaged across both management options.   Although, 37% of participants changed their WTP amount between MS1 and MS2 (20% upwards; 17% downwards), mean WTP was not significantly different, rising from £3.67 to £4.49. Comparison with the survey group reveals that WTP is significantly lower in both the MS groups, with a mean WTP from the survey of £15.90. Hence the survey based CV produced an estimate of mean WTP that was 3.5 times higher than that generated by the Market Stall approach.

5.5.13   Figure 5.9 provides estimates of mean WTP for Projects C and D. WTP for Project D was higher in MS1, MS2 and the main survey. This was expected *a priori* as Project D protected all four species, whereas Project C protects only the two endangered species. However, the only statistically significant difference between options arises from MS2. This

---

[1] This is the same category used in the results section of the CV survey described in Section 5.2

could suggest that participants, by the end of MS 2, were better able to differentiate between the two policies.

Figure 5.8   Mean WTP for MS1, MS2 and the Main CV Survey

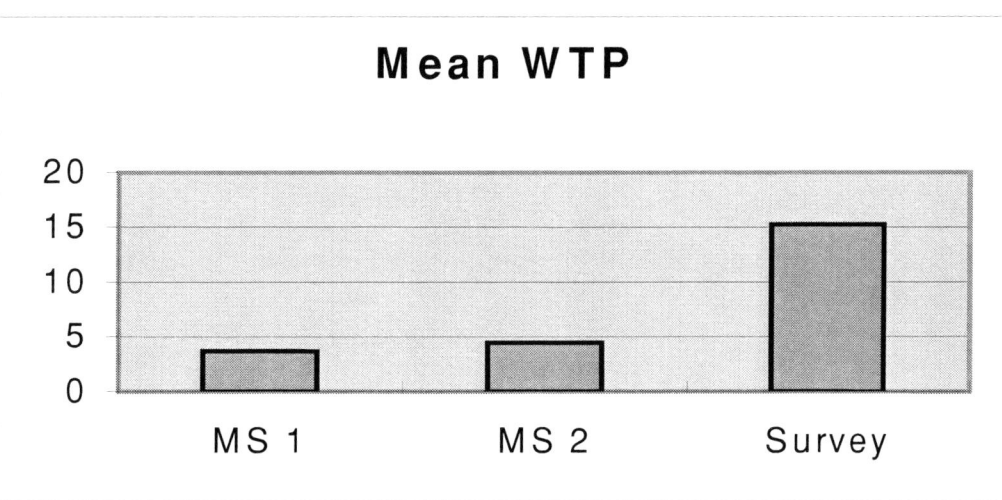

Figure 5.9   Mean WTP for Options C and D for MS1, MS2, and Main CV Survey

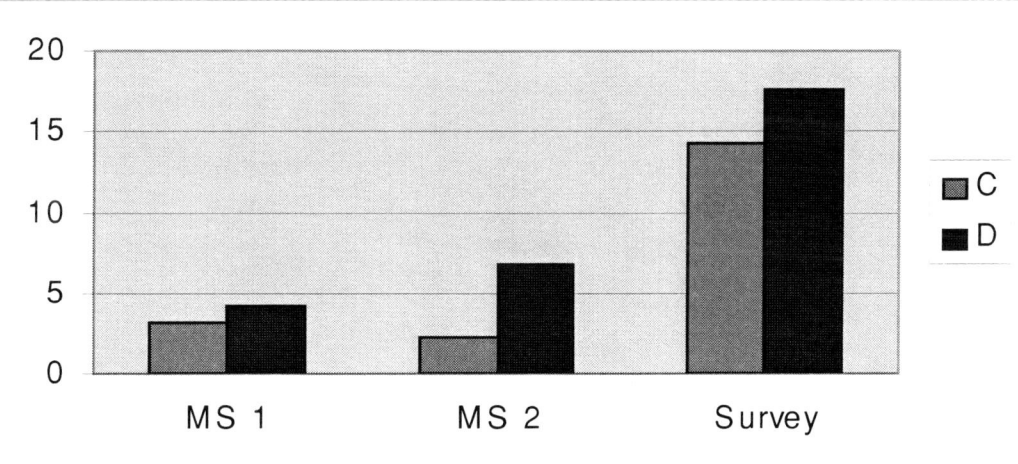

5.5.14 Another test of validity is to establish whether WTP estimates can be predicted from socio-economic and attitudinal variables in regression analysis. The results of both a TOBIT and an OLS analysis confirm that MS WTP, just as in the main survey, was significantly correlated with variables we would expect to influence WTP. For example, participants who were on higher incomes *(income)*, were members of environmental groups *(member)* and ranked wildlife protection *(wildlife)* and goose conservation *(goose)* higher as a priorities had higher WTP. The dummy variable for the management option (C or D) was also found to be significant, confirming that once differences in the samples controlled for in the regression have been taken into account, mean bids for Option C were *significantly* lower than bids for Option D.

5.5.15 An interesting finding was that the adjusted R^2 value, which measures the degree to which the variability in individual bids can be explained by the independant variables, was much higher for the MS2 data (adj. R^2 = 0.34) than for the survey data (adj. R^2 = 0.18). This supports the argument that mean WTP estimates derived from the MS are more reliable than those derived from the survey.

5.5.16 The MS also performed well in relation to protesting. Only 4% of MS participants were classified as Protesters compared to around 30% in the survey. One explanation for this could be that the discussions within the MS groups helped to counter notions that might lead to protesting (for example negative views about taxation could be reduced by explaining that tax was the only way to pay). In a survey there is less opportunity to persuade respondents of the case for using taxation. Another possibility is that 'protesting' is viewed as an easier option than having to consider the details of the CV option by respondents who are anxious to 'escape' from the interview situation.

5.6   Aggregation

5.6.1   In all applications of environmental cost-benefit analysis, we are interested in trying to make informed guesses about what the economic benefits to the population as a whole of a policy or project are, based on the information gained on values held by those people whose preferences we have sampled. In the context of this study, the most relevant question to ask would seem to be what the likely benefits are to the Scottish population as a whole, since the geese payments scheme comes from the Scottish Executive budget.

5.6.2   Taking into account the fact that estimates of WTP from CV surveys often over-state the amount people really would pay by a considerable margin (Macmillan, 1999) and the findings of the MS experiment it was decided to calibrate WTP estimates from the main CV survey downwards by dividing by 3.5. Using the most recent figure of 2,186,500 households in Scotland, the annual, aggregate benefits of the four policy options are given in Table 5.7.

Table 5.7: Aggregate benefit estimates for policy options A-D			
Option	Mean WTP CV Survey (£/household/year)	Calibrated Mean WTP (£/household/year)	Total Benefits (year)
A: no shooting	8.32	2.38	5,203,870
B: avoid 10% fall	10.99	3.14	6,865,610
C: obtain 10% rise endangered species	16.28	4.65	10,167,225
D: obtain 10% rise all species	20.97	5.99	13,097,135

5.6.3   Clearly, these are very large numbers, but this is inevitable given the assumed size of the relevant population. Multiplying even a small per-household value by over two million will give a very big value.

5.7   Conclusions

5.7.1   What conclusions can be drawn from the benefits study? First, in terms of general attitudes, we found that people rate wildlife protection as a relatively important component of rural policy. Wild geese conservation is rated less important than most other conservation issues raised with the public. Despite this, there was clear majority support for geese conservation policy, even when this is costly.

5.7.2   The willingness to pay surveys found that different attributes of goose conservation policy were valued differently by the various groups. For instance, the general public, and Islay and Strathbeg visitors were willing to pay between £9, £7 and£19 per household per year respectively for a management option that did not involve shooting. Residents, on the other hand, were not willing to pay anything for this type of policy.

5.7.3   The qualitative and quantitative research strongly suggests that all groups favoured conservation policies that target endangered species. Statistical analysis of the willingness to pay results showed that the general public were not prepared to pay significantly higher levels of additional taxation for polices that extended conservation measures to non-endangered species. Both visitors and local residents were actually prepared to pay **more** for a policy that included endangered species only, than one that included all species.

5.7.4   There was some evidence that people would support small increases in goose numbers (and avoiding a small fall) but would not pay for more substantial increases in goose numbers. For example, local residents at Strathbeg and on Islay, said they would be willing to pay £13 per household on average to avoid a 50% increase in the goose population.

5.7.5   Choice experiments are particularly good at estimating attribute values for environmental goods, in other words in decomposing the value of (in this case) a policy into its constituent parts. Contingent valuation is better at estimating discrete changes in a 'policy package' that includes a number of attributes. For example, a 10% increase in geese numbers achieved through no shooting and on at all locations in Scotland. The main drawback to CV is that it would be very expensive to use it to look at the values of all attributes and their levels as part of policy design. CE and CV can thus be seen as serving somewhat different functions in the policy appraisal process.

5.7.6   There was some good validatory evidence that the CE and CV surveys performed well in relation to current standards. However, results from both the feedback research and the MS experiment suggests that the survey approach is not entirely satisfactory when valuing unfamiliar environmental goods. As the MS approach allows respondents more time and information than is possible under normal survey conditions and the opportunity to discuss the valuation question with other household members, it may have the potential to produce more reliable estimates of WTP.

CHAPTER SIX   AGRICULTURAL COSTS

Summary

This chapter describes the main results of a study into the economic costs of goose damage to farming.

The research aims to understand how geese affect farming activity and to estimate the total and marginal economic costs of wild geese by grazing.

The research comprises a review of previous estimates of goose damage and associated costs, and original survey work in the two case study areas:

- Islay: a predominately livestock farming area with high goose densities
- Strathbeg: a predominantly cereal cropping area with low to moderate goose densities.

On the next page the principal results of the research are highlighted.   More detailed information on data collection, analysis and assumptions used is available in Technical Report C. (A copy of the farm questionnaire is also included in the Technical Report).

Key Findings

- The 1999/00 estimated total annual cost of goose damage on Islay and at Strathbeg was, respectively, £560 000 and £220 000, with an average cost per farm surveyed of £11 500 and £5 800.

- The higher costs on Islay reflect the higher population density of geese and the higher proportion of available land that is affected both "per farm" and in aggregate.

- At Strathbeg much of the damage is to cereals, whereas on Islay, where nearly all the grass area is affected by geese, damage to the grass/livestock sector accounts for 93% of all costs.

- At both locations, rising goose numbers lead to greater economic costs to farmers, but the Islay survey suggests that the marginal costs of geese damage decline rapidly as goose densities rise to higher levels.   On Islay, where the geese are concentrated on a relatively small island, this may be due to strong competition amongst the geese for limited grazing. Although these diminishing marginal effects are not marked at current <u>average</u> goose numbers and densities, this finding suggests that if numbers continue to rise in future, a situation may be reached where Islay will approach its capacity to accommodate increasing numbers of geese.

- At Strathbeg and other locations on the mainland where geese can easily move to other areas. At these lower densities there is less competition and hence the marginal costs of damage decline less.

6.1 Introduction

6.1.1 The research aims to understand how geese affect farming and to estimate the total and marginal economic costs of wild geese grazing. The research focuses on two case study areas: Islay, a livestock farming area with high goose densities, and Strathbeg, a predominantly cereal cropping area with much lower goose densities.

6.1.2 In order to meet these aims, both qualitative methods (Focus Groups) and a detailed farm survey involving in-depth interviews with a sample of farmers were used. A farm survey approach has never been used before on Islay as aggregate costs have always been predicted from physical parameters of damage which were priced and then grossed up to the whole island. This modelling approach implicitly assumes a uniform relationship between damage and goose numbers and hence that all farms are affected in the same way.

6.1.3 A farm survey was considered preferable because the effects of geese can be investigated for individual farms. Also, additional information on the effects of geese could be gleaned from discussions with farmers and, more importantly, some insight gained into the relationship between costs and goose numbers.

6.1.4 Prior to the survey an extensive review of relevant research was undertaken. A number of studies which throw light on the damage and extra costs caused by geese to Scottish agriculture were identified and the following conclusions were drawn:

- there are significant losses of crops and grassland to geese grazing
- damage is variable between seasons, locations, timing and grazing pressures
- most scientific studies of goose damage have been partial and have not examined the effect of goose grazing on the whole farm.
- where whole farm studies have been undertaken they have indicated average costs due to geese of between £17 and £29 per hectare (£3000 to £8000 per affected farm).
- it has been difficult to identify the marginal economic costs of geese.

6.2 Focus Group Research

6.2.1 The focus groups were intended to inform the consultants about the impacts of geese on agricultural activities in both locations prior to the implementation of the survey. A focus group was held at each location and provided excellent information on the relative importance of different types of goose damage and trends in the timing and numbers of geese. In addition, the participants freely expressed opinions on, and attitudes to, a range of goose-related issues such as: scaring and the effects of geese on farming systems and the local economy.

6.2.2 Two focus groups were held for farmers around Strathbeg and on Islay with 15 farmers in total participating. The Strathbeg group farmed 3350 ha (mostly crops & grass) as owner-occupiers, whereas the Islay group farmed 2800 ha acres (predominantly rough grazing) mostly as tenants.

6.2.3 The main findings from the focus groups are summarised below (a fuller commentary for each location is included in Technical Report C).

- *Distribution of geese*
The geese graze near the roost on arrival and before leaving in the spring. Around Strathbeg, geese penetrate further inland during mild winters. On Islay there are multiple roosts so spread is more difficult to define.

- *Types of damage*
The grazing damage around Strathbeg is mostly to winter cereals and early grass. This leads to late turnout of stock, loss of early bite for sheep, late sowing of spring barley, late nitrogen application and reduced yields. Little damage is caused to swedes, potatoes, OSR and autumn stubbles around Strathbeg. On Islay the loss of winter grazing and early bite is significant and silage is delayed. Winter cereals are not grown because of the geese, puddling is a major factor, and grass needs reseeding more frequently. Whole crop silage is not generally damaged and is replacing second cut silage

- *Severity of damage*
The main factors affecting the degree of damage in both Islay and Strathbeg are location and weather. Geese have strong field loyalty and avoid fields near busy roads, buildings and pylons. Weather not only affects the dispersion of the geese but also the recovery of the damaged crops and grass. Other factors include degree and intensity of scaring, timing of grazing in the growing season and, on Islay, ESA regulations.

- *Goose numbers and damage*
There were large increases at Strathbeg in geese numbers during the 1980s, with smaller increases in the 1990s although the amount of damage is now related more to growing conditions rather than numbers. On Islay, there has been a 30% increase in geese over the last 10 years, resulting in more areas affected and increases in the severity of the damage. Around Strathbeg the relationship between numbers and damage was thought to be linear though some believed that marginal damage increased with goose numbers. On Islay the relationship was also perceived to be linear or worse. In both locations, it was felt that geese were becoming bolder and less frightened of people, buildings and scaring.

- *Changes in the farming system*
System changes brought about by the geese included: no winter cereals in fields favoured by geese, late sowing of spring cereals, need for extra winter keep and supplementary feeding. On Islay, the geese also forced farmers to reseed more frequently and to grow whole crop silage. No obvious change in land values was perceived at either location. However, it was felt that the presence of geese might reduce grass let prices.

- *Benefits of the geese*
Around Strathbeg, farmers felt that the main benefit of geese to farming was cleaning up the stubbles and crop leavings in autumn. Any economic gain resulting from visitors to the area tended to accrue to the tourist industry and to goose guides in the area who shot over their land with wildfowlers in the open season. Some farmers

received free scaring services from the guides in the close season. On Islay there were no perceived benefits to agriculture but most (though not all) farmers received the IVGMS payments and a small number also received SSSI payments for geese.

- *Future management*
The farmers from Strathbeg recognised that increased shooting and the sale of goose meat was probably infeasible. They appreciated the mid-1990s SNH-funded scheme and were hoping for a new scheme as they had adapted their systems as far as they could. On Islay, the farmers felt that the only solution was to maintain or increase payments. Islay participants were not very aware of the new proposals but hoped that a new scheme would be at least as attractive as IVGMS. However, they were critical of the skewed distribution of payments under IVGMS, and some were critical of the reliability of the SNH counts.

6.3   Farm Survey

6.3.1   The focus group findings aided the design of the farm survey questionnaire which aimed at quantifying (in economic terms) the losses identified in the focus groups and the costs of mitigating the effects of geese.   In addition, the survey included a number of qualitative questions on trends, the net effects of increasing goose numbers and how farmers may cope with "the goose problem" in future.   The same questionnaire was used at both locations.   The survey was based on in-depth interviews of a limited number of farmers: 18 on Islay and 15 at Strathbeg.   It was conducted within two months of the focus groups and included a number of focus group participants at each location.   The sample was chosen on the basis of a spread of distances from the roosts, different sizes of farm and types of land tenure.

6.3.2   The basis for estimating the costs of geese was farmers' recall of the situation during the winter of 1999/00.   Farmers were asked technical questions such as the number of cattle and the weeks delay in turning out last spring due to the effects of geese, and the winter rations fed, whilst the consultants calculated the costs of delayed turnout using standard prices for feeds.   For the effects on silage, the loss of yield and the number of weeks delay in cutting were obtained, and the consultants priced the silage losses and the value of reduced aftermath grazing.   For arable, farmers estimated reduced cereal yields due to grazing of winter crops or late planting of spring cereals, and the economic consequences were calculated separately using standard prices.

6.3.3   This approach, of asking farmers questions on the physical effects of geese and then translating these into economic costs, was also used for the other anticipated components of goose effects.   This included the opportunity costs of a changed farm system to accommodate geese, and the effects of more frequent reseeding of grassland.

6.3.4   Although the farmers' responses were the basis for calculating the effect of geese, some adjustments at individual farm level were necessary where:

- the suggested impact of geese was not credible in relation to other sources of information
- unreliable assumptions were being made by farmers

- farmers had difficulty envisaging a "no geese" situation (many farmers, especially on Islay, had operated farm systems adapted to geese for all their working lives).
- the real effects of geese had to be 'teased-out' (e.g. other factors can affect timing and stocking rates, such as ESA conditions for other birds (e.g. corncrakes) and weather conditions.
- double counting was a problem (e.g. between loss of early grazing and delayed silage making).

6.3.5   At the end of each survey, results were carefully examined by the consultants for anomalies through cross-referencing against earlier work and with standard data sources such as the Farm Management Handbook.   The results were then raised using Agricultural Census data, and in the case of Islay where individual goose counts were available, a statistical analysis was conducted to relate costs to goose populations and densities.

6.4   Islay Damage Costs

6.4.1   The 18 farms sampled on Islay extended to 14,600 ha and accounted for 38% of the productive land on the island and 37% of the 1997/8-1999/00 goose population (an average density of 5.7 per ha, close to the island's overall density). On average, these farms extended to 159 ha of productive land (rotational and permanent grassland and crops).

6.4.2   Total costs for all 18 Islay farms in the sample were £206,000 or an average of £11,500 per farm (Table 6.1 and Figure 6.1). Main findings are

- Loss of early grazing was the most costly goose effect, followed by losses to silage (including aftermath and hay)
- Including the relatively low cost of additional grass fertiliser (typically applied in the spring), some 67% of all costs measured were accounted for by losses to grass in the first part of the season.
- Additional reseeding costs and losses of winter grazing were also significant, but direct losses to cereal crops were relatively unimportant.

Table 6.1: Economic Costs of Geese – Islay Sample*				
Type of Goose Effects	Total Economic ('000) Cost £	Average/farm ('000) £ (18 farms)	Costs/prod. ha £ (2855 ha)	Costs/goose £ (16213 geese)
Early Bite	77	4.3	26.9	4.74
Winter grazing	20	1.1	7.0	1.23
Silage and aftermath	54	3.0	18.9	3.33
Extra Grass Fertilisers	85	0.5	3.0	0.52
Extra Reseeding	33	1.8	11.6	2.05
Total Grass costs	*192*	*10.7*	*67.3*	*11.86*
Crop Losses	9	0.5	3.2	0.57
Winter Barley not grown	5	0.3	*1.8*	*0.31*
Total Arable costs	*14*	*0.8*	*5.0*	*0.88*
ALL costs	206	11.5	72.3	12.74

* rounded to the nearest thousand.

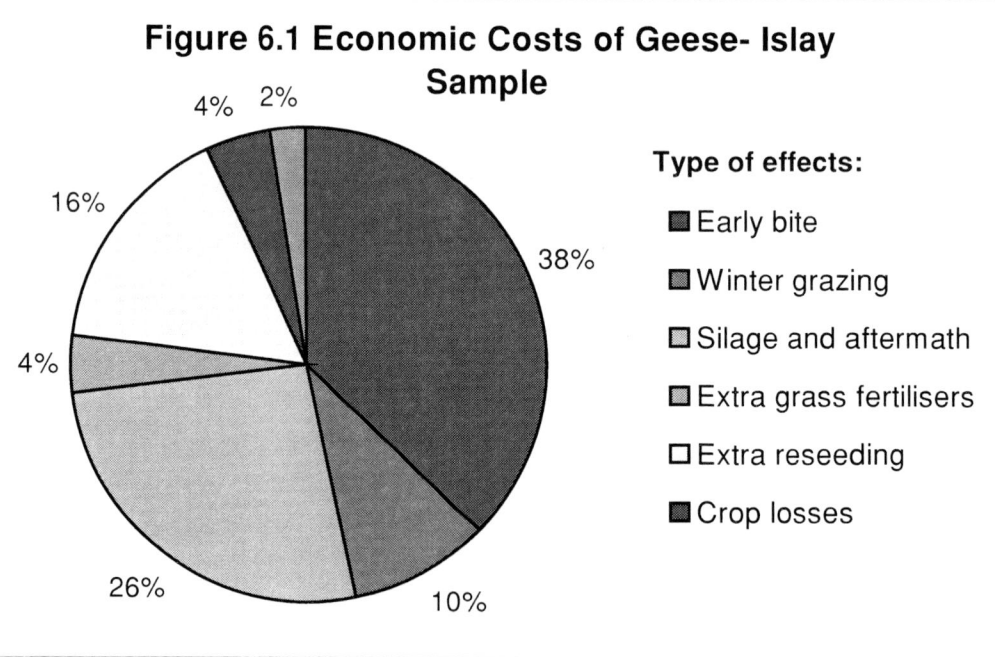

Figure 6.1 Economic Costs of Geese- Islay Sample

Type of effects:
- Early bite
- Winter grazing
- Silage and aftermath
- Extra grass fertilisers
- Extra reseeding
- Crop losses

6.4.3 In order to estimate the goose-related costs incurred by all farms on Islay, a figure of 2.703 was used to raise the sample results – based on the 37% share of Islay productive land and goose numbers included in the 18-farm sample. This indicated an Islay-wide cost figure of £560,000 for 1999/00 for those items included in the survey. This is some £20,000 less than the implied costs estimated by the Islay Goose Management Group in its proposal to the

National Review Board in June 2000 and mid-way in Lilley's 1997 range of costs, although completely different approaches were used. The estimates are higher than those calculated by Westmacott, (2000) for a sample of 14 farms in goose-affected areas of Kintyre. Average costs in this study were £7,200 per farm or £29/ha.

6.4.4   In addition, farmers indicated in both of the focus groups and in the survey that there are other effects of geese not covered by the survey (nor any previous study). These include: reduced lambing rates, later lambing and calving, and an overall reduction in breeding livestock than would occur in a "no geese" situation. Since it is difficult to value these impacts accurately and uniformly across all farmers in the survey, these impacts were not estimated.

6.4.5   The availability of SNH's goose population data for individual farms on Islay (for calculating payments to farmers since 1992) enables some analysis to be made of the relationship between goose numbers and densities, and economic costs, at the level of individual farms in the survey.

6.4.5   The most interesting relationship is that relating the total of all economic costs measured per farm to the average no. of geese per farm (1997/8 to 1999/2000). The best fit (based on adjusted R^2) was obtained from a quadratic relationship that indicates diminishing marginal costs. Using the quadratic relationship as a predictor, the marginal economic costs per goose decline from £12.52 at low populations to less than a £1 at high populations (Table 6.2 and Figure 6.2). The mean situation for the whole sample was 900 geese per farm and £12.74 costs per goose (weighted average).

Table 6.2 Predicted Marginal Costs against Geese Numbers and Density (Islay only)		
No. Geese per farm	Total Costs per Farm (£) *	Marginal Costs per goose (£)
500	7700	-
1000	14,000	12.5
1500	18,800	9.5
2000	22,000	6.6
2500	23,700	3.3
3000	24,000	0.5

* $A = 16.9923\,D - 0.0030\,D^2$ where A = costs per farm, D = geese per farm

Figure 6.2: Relationship between goose density and marginal cost

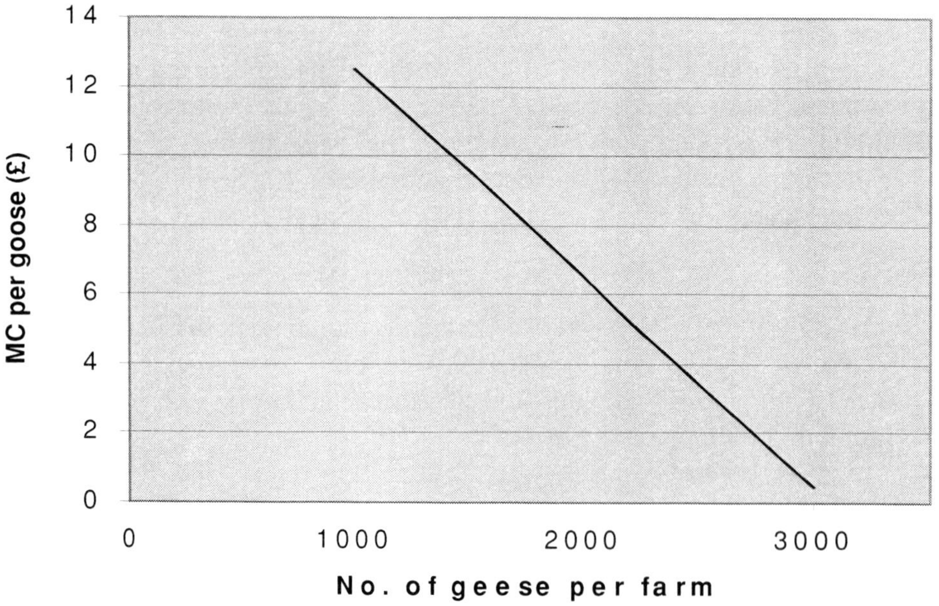

6.4.7   In practical terms, this relationship suggests that, for a fixed area, with a fixed amount of herbage available as on Islay, the more geese which graze, the less will be the consumption per head (due to increased competition) and consequently the lower the damage per head. Another possible explanation is that the proportion of barnacle geese increases at higher densities. As these geese are smaller and eat less per bird than the larger White-fronted goose, then it is expected that there will be less damage per goose at higher densities.

6.4.8   The above analysis was based on 3-year average goose numbers recorded by SNH in seasons 1997/8, 1998/99 and 1999/00. A further analysis was carried out for goose numbers relating only to the most recent figures (1999/00). Similar results were obtained and similar conclusions drawn about the relationship between goose density and marginal costs.

6.5   Strathbeg Damage Costs

6.5.1   The Strathbeg "catchment" was represented by 15 farms covering 4,900 ha of which 4,500 ha were crops, grass and set-aside, and which accounted for 33% of the estimated area of significant goose grazing. Their average size was 272 ha of crops and grass (71% greater than the Islay sample). Using a fairly crude estimation of the goose population in the Strathbeg sample these 15 farms experienced an overall density of 1.5 geese per productive ha last winter (25% of the density on Islay, but 36% higher than the density for all farmers in the Strathbeg catchment)

6.5.2   Table 6.3 and Figure 6.3 provides an estimate of the economic costs of geese for the 15 farms of the Strathbeg sample. Total cost for the sample was approximately £88 000 with an average of £5,800 per farm, and £22 per productive hectare (excluding rough grazing and set-aside).

Table 6.3 Economic Costs of Geese – Strathbeg			
Type of Goose Effects	Total Economic Cost £ ('000s)	Average/farm £ (15 farms) ('000s)	Costs/prod ha £ (4080 ha)
Early Bite	15	1.0	3.6
Winter grazing	0	0	0.0
Silage and aftermath	4	0.2	0.9
Extra Grass Fertilisers	1	0.1	0.3
Extra Reseeding	0	0	0.0
Total Grass Costs	**20**	**1.3**	*4.8*
Crop Yield Losses	32	2.1	7.8
Resowing crops	2	0.1	0.4
Extra crop inputs	2	0.1	0.4
Total crop costs	**36**	**2.3**	*8.6*
WB not grown	33	2.2	8.0
Total arable costs	**69**	**4.5**	-
ALL damage costs	**89**	**5.8**	**21.5**

Figure 6.3   Economic Costs of Geese- Strathbeg sample

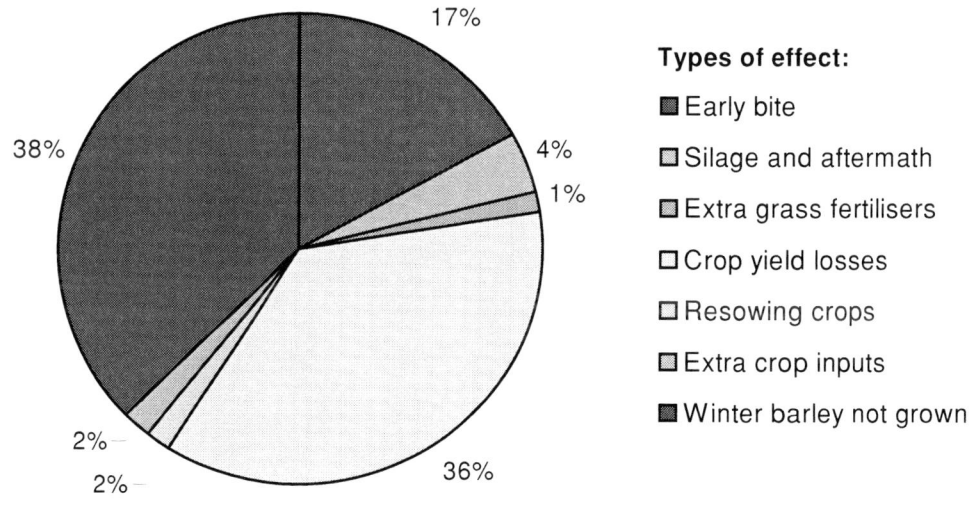

Types of effect:

◼ Early bite

▢ Silage and aftermath

▨ Extra grass fertilisers

☐ Crop yield losses

☐ Resowing crops

▨ Extra crop inputs

◼ Winter barley not grown

6.5.3   Main findings are

- costs are predominantly arable-related (40% cereal losses and additional inputs to cereals, and 37% relate to the effect of goose grazing in reducing winter cereal areas)
- Grass-related costs (22% of all costs) were dominated by the effect of lost early bite on cattle feeding
- Silage losses are relatively unimportant and the costs of lost winter grazing are zero.

53

6.5.4   These results reflect the predominance of arable-beef systems where there are large areas of stubbles and good supplies of straw for winter feeding. Compared with Islay, the costs per farm are 50% less, and the costs per hectare are only 30% of the Islay level.   This can mainly be attributed to the lower goose densities at Strathbeg.

6.5.5   Raising the Strathbeg sample results to the total area affected by geese roosting on the Loch of Strathbeg is more difficult than on Islay because there are no discrete boundaries of goose activity and the grazed areas do not correspond to whole parishes. However, using 10-year old data on the spatial aspects of goose densities (Keller *et. al* 1997), it was possible to make a rough estimation of the main area of goose grazing and to relate this to the area covered by the 15 farms.

6.5.6   The sample accounted for 33% of the crops and grass area of the "catchment" of Strathbeg geese and 40% of the total goose population. A factor of 2.5 was used to raise the sample results to the whole area vulnerable to goose damage.   The economic costs for the whole area are therefore **£219,000**. (As in the case of Islay, some additional costs cited by farmers could not be quantified such as later lambing).

6.5.7   The Strathbeg survey data do not permit any reliable analysis of the effect of goose densities on damage, but each farm was classified as "Low", "Medium" and "High" on the basis of proximity to the roost and the Keller *et al.* (1997) mapping exercise of 1990 (Table 6.4). Although the analysis provides a clear indication that economic costs rise with increasing goose densities it is not possible to deduce whether the relationship is linear, or one of diminishing marginal costs.

Table 6.4 Relationship between goose density and damage costs at Strathbeg				
	Low	Medium	High	All
No. of farms	4	5	6	15
Productive ha	1,328	906	1,845	4,080
Economic Costs £	7,727	18,355	61,443	87,525
Costs per productive ha £	5.8	20.3	33.3	21.5
Economic Costs per farm £	1,932	3,671	10,241	5,835

6.5.8   Scaring costs for the Strathbeg catchment are around £17 000 per year but these are mainly offset by £10-15 000 of benefits generated by shooting in the open season (Benefits based on the revenue from shooting lets and the value of 'scaring' services provided by goose guides).

6.5.9   The results from this study appear to be consistent with previous studies. Daw's study at Strathbeg (1992) covered 7100 ha on 24 farms and estimated total goose-related costs at £184,000 or £7,700 per farm (£26 per ha).   A similar investigation of 42 farms covering 7700 ha around Loch of Skene (de Otyeza, 1995) found costs of £130,000 or £3100 per farm (£17/ha).

6.6   Comparison of main findings: Islay and Strathbeg

6.6.1   Table 6.5 summarises the main findings for both Islay and Strathbeg, and compares them in terms of areas affected, goose densities and costs both for the sample farms and for the whole "catchment" at each location. The raising of the survey results for Strathbeg is less reliable than for Islay because there are no goose counts available for individual farms at Strathbeg.

Table 6.5 Comparison of main findings: Islay and Strathbeg		
	Islay	Strathbeg
Sample:		
Area productive land (ha)	2856	4469
Area grass (ha)	2551	1867
% of grass area affected by geese	95%	54%
% of cereal area affected by geese	75%	30%
No. geese (000)	16.2	6.6†
Geese per prod. Ha	5.68	1.48†
All economic costs* of sample (£000)	206.5	87.5
Costs per farm (£000)	11.5	5.8
Costs per ha productive land excl. set aside (£)	72.3	21.5
Costs to grass (£000)	192.3	19.6
Costs per grass ha (£)	75.4	10.5
Costs ** to crops (£000)	14.2	67.9
Crop related costs **per ha of existing crops (£)	46.6	30.7
All economic costs* per goose £	12.74	13.26†
Total scaring costs ('000)	0	6.8
"Catchment"		
Total area land (ha)	56,881	15,572
Total area productive land in catchment (ha)	7570 (13%)	14,342 (92%)
No. geese in "catchment" (000)	44.3	15.6
Geese per productive ha	5.85	1.09
Estimated costs* for "catchment" raised from		
Figures from survey (£000)	558	219

* excluding scaring costs   ** includes opportunity costs of reduced winter cereals.
† significant margin of error attached to geese numbers for Strathbeg sample.

6.6.2 Goose populations and densities are much higher on Islay and affect higher proportions of the available land. This leads to greater amounts of damage and other costs on Islay, both "per farm" and in aggregate. However, the costs per goose are remarkably similar at each location. The economic effects of geese on crops are far greater at Strathbeg because of the much larger area of cereals. On Islay, nearly all the grass area is affected by geese, and damage to the grass/livestock complex accounts for 93% of all costs.

6.6.3 At both locations, rising goose numbers lead to greater economic costs to farmers, but the survey suggests, for Islay, that the incremental costs of geese decline as numbers rise to high levels. This may be due to the fact that, at high densities, there is insufficient herbage and that competition amongst the geese results in less consumption per bird. Although these diminishing marginal effects are not marked at <u>average</u> goose numbers and densities, this finding suggests that if numbers continue to rise in future, a situation may be reached where economic costs reach an absolute maximum. On some Islay farms with very high densities of geese, this point may have been reached already.

6.6.4 It has not been possible to analyse this marginal aspect at Strathbeg. However, it seems likely that the relationship between goose numbers and costs is linear at current population levels since the Strathbeg geese can extend their geographical range or even graze some of the areas relatively near to the roost which are currently unaffected. At Strathbeg some 30% of the cereal area on the survey farms was affected by geese, compared with 75% on Islay. For grassland the figures were 54% and 95% affected, respectively.

6.7   Conclusions

6.7.1   The following conclusions are drawn from the Focus Group discussions, the opinion survey section of the survey questionnaire, and especially from the estimates of goose-related costs derived from the survey and subsequent calculations.

- The economic losses due to wild geese at farm level measured in this study are significant. On Islay costs average £11,500 per affected farm, and £5, 800 at Strathbeg. The estimated total costs of wild geese in the whole case-study area is estimated to be £560,000 on Islay and £236,000 at Strathbeg.

- Most of the Islay costs are to grassland especially the loss of early grazing, reduced and late silage, and the additional costs of more frequent reseeding. At Strathbeg, the main economic effects of geese are related to arable cropping, particularly reduced cereal yields and a reduced ability to safely grow winter barley. The loss of early bite is also important at Strathbeg where significant numbers of cattle are necessarily turned out late in the spring.

- On Islay and at Strathbeg, there is strong evidence of increasing levels of cost with rising goose densities. However, on Islay, where there were good data on goose numbers for individual farms, the costs per goose fell as density increased. At high densities, the effect of still more geese is quite small, presumably because competition for limited grass is increased and consumption per goose is reduced. This has possible implications for the effect on costs of changing goose populations on Islay in future, and the appropriate levels of compensation paid to farmers.

- At Strathbeg, it was not possible to assess the existence of diminishing marginal damage. However, it seems likely that a more linear relationship occurs between goose numbers and costs because very high densities are not achieved due of the greater dispersal of the geese.

- The results presented here reflect the costs associated with geese in one particular season. Due to seasonal variation in growing conditions, goose numbers, and in agricultural prices it is to be expected that these estimates will vary from year to year. Furthermore, it is worth the reminding the reader that the estimates provide are based on a relatively small sample of farms affected, hence there is likely to be some error in relation to grossing up the costs in both case study areas.

CHAPTER SEVEN   OVERALL CONCLUSIONS AND RECOMMENDATIONS FOR FURTHER RESEARCH

7.1   Conclusions

1   In terms of general attitudes, we found that people rate wildlife conservation as a relatively important component of rural policy. Wild geese conservation is rated less important than most other conservation issues raised with the public. Despite this, there was clear majority support for geese conservation policy, even when this is costly.

2   The willingness to pay surveys found that different attributes of goose conservation policy were valued differently by the various groups. For instance, the general public and visitors were willing to pay between £7-£19 per household per year for a management policy that did not involve shooting. Residents, on the other hand, were not willing to pay anything for this type of policy.

3   The qualitative and quantitative research strongly suggests that all groups favoured conservation policies that target endangered species. Statistical analysis of the willingness to pay results showed that the general public were not prepared to pay significantly higher levels of additional taxation for polices that extended conservation measures to non-endangered species. Both visitors and local residents were actually prepared to pay **more** for a policy that protected endangered species only, than one that included all species.

4   There was good evidence that most people would be willing to pay for policies that encouraged a small increase (10%) in endangered species, but there was also some evidence that people would not pay for more substantial increases in goose numbers. For example, local residents said they would be willing to pay £13 per household on average to **avoid** a 50% increase in the goose population.

5   The overall total annual willingness to pay of the Scottish population for alternative goose management policies ranged from £5.2 million for a policy that does not rely on shooting geese to £13.1 million for a policy that achieved a 10% rise in all migratory geese species.

6   CE is particularly good at exploring the values placed on individual attributes of a particular policy: for example, a 10% increase in geese numbers alone. CV is better suited to estimating WTP for discrete changes in policy packages: for example, a 10% increase in geese on Islay achieved through habitat management alone. CE and CV can thus be seen as serving somewhat different functions in the policy appraisal process.

7   The estimated total costs of goose damage on Islay and around Strathbeg in 1999/00 were £560 000 and £220 000, with an average cost per farm surveyed of £11 500 and £5 800 respectively. The higher costs on Islay reflect the higher population density of geese and the higher proportion of available land that is affected both "per farm" and in aggregate.

8   At both locations, rising goose numbers lead to greater economic costs to farmers, but the survey suggests for Islay that the marginal costs of geese damage decline rapidly as numbers rise to high levels. On Islay, where the geese are concentrated on a relatively small island, this may be due to strong competition for grazing amongst the geese at high densities.

In Strathbeg and other locations where geese can easily move to other areas, there is less competition and hence the marginal costs of damage decline less.

9   Although these diminishing marginal effects are not marked at **average** goose numbers and densities, this finding suggests that if numbers continue to rise in future, a situation may be reached where Islay approaches its capacity to accommodate increasing numbers of geese.

10   The agricultural costs associated with geese reported in this study refer to one particular season. Due to seasonal variation in growing conditions, goose numbers, and in agricultural prices it is to be expected that these estimates will vary from year to year. Furthermore, it is worth the reminding the reader that the estimates provide are based on a relatively small sample of farms affected, hence there is likely to be some error in relation to grossing up the costs in both case study areas.

7.2   Recommendations for Further Research

1   Further investigation of the marginal benefits of increasing the wild goose population is merited. The CE provided some interesting insights into this issue but failed to establish consistent or significant WTP estimates for future changes in the wild goose population amongst the general population.

2   No attempt was made to estimate the willingness to accept (WTA) compensation of people who had negative views toward the conservation and management of wild geese. Although the majority of people supported the policies investigated in this study, a minority of respondents, particularly in the case–study areas were opposed. Incorporating their WTA compensation may substantially reduce the net benefits of wild goose conservation in Scotland (see for example Macmillan and Duff, 1998)

3   The MS experiment has suggested that giving people more time to think about the issue and consider their preferences results in lower and possibly more reliable CV WTP values. It would be very useful to extend this research to a wider selection of projects and a more representative sample of respondents (e.g. local residents). It would also be interesting to attempt a similar exercise with the CE method.

4   More farm survey work at both sites could be justified using larger samples than in the current survey, in order to provide greater (or less) substantiation of the conclusions about the relationship between damage and goose numbers. With appropriate data collection and monitoring of the new Islay GMS at farm level, further useful research could be carried out on how goose densities affect damage.

5   To assist this research it would be very useful to have more frequent and more reliable information on the location and density of geese in affected areas. This is particularly true at Strathbeg where the most recent spatial analysis was conducted ten years ago.

6   In future, a simplified method for assessing economic costs may be needed on a year to year basis. One approach would be to record selected key indicators of damage (turnout dates, silage timing, resowing, cereal yields etc) on a regular panel of selected farmers. Such records could be supplemented by other measurements carried out externally, e.g. sward

damage along transects, goose dropping counts or subjective assessments of damage by agricultural specialists. These indicators could then be weighted into an index of economic costs for inter-year comparison. This index would need calibration from a baseline survey of costs on the panel of farms possibly using similar approaches to those employed in this study.

LITERATURE CITED

Daw D., 1992. The Economic Impact of Grey Geese on Agriculture around the Loch of Strathbeg. *MSc dissertation (unpublished) University of Aberdeen.*

de Oteyza E. S. G., 1995. Impact on Agriculture of the Lower Donside caused by Greylag Geese roosting at the Loch of Skene. *Undergraduate dissertation (unpublished) University of Aberdeen.*

DETR (2001) *Guidance on Using Stated Preference techniques for the Economic Valuation of Non-Market Effects*". Department of the Environment, Transport and the Regions, forthcoming.

Feare C., 1990. Agricultural Conflict and Licensing in England and Wales. In Owen & Pienkowski, *The damage-conservation interface illustrated by geese.* Ibis 132.

Hanemann, W.M. 1994. Valuing the environment through contingent valuation. *Journal of Economic Perspectives 3, 1-23.*

Hanley N, Mourato S and Wright R (2001) "Choice experiments: a superior alternative for environmental valuation" *Journal of Economic Surveys*, forthcoming.

Hanley N and Spash C (1993) *Cost-Benefit Analysis and the Environment*. Cheltenham: Edward Elgar.

Islay Goose Management Group, 2000. Outline Proposals for IGMS – submission to National Review Body.

Keller, Gallo,-Orsi, Patterson, Naef-Daenzer, 1997. Feeding Areas of Pinkfoot Geese, Strathbeg. Wildfowl 48

Lilley R., 1997. Greenland Geese Wintering on Islay – Assessing the Agricultural Impact. *SNH*

Macmillan, D.C., Smart, T. S. and Thorburn, A. P. 1999. Validation of the Contingent Valuation Method: A comparison of real and hypothetical donations to an environmental Trust. *Environmental and Resource Economics* 14(3): 399-414.

Macmillan, D.C. and Duff, E.I. 1998. The non-market benefits and costs of native woodland restoration. *Forestry*, 71(3): 247-259.

Marston, A.D. 1998 Tourism and other economic benefits associated with wild geese on the isle of Islay, Scotland. . *MSc dissertation (unpublished) University of Aberdeen.*

RSPB/BASC. 1998. Geese and local economies in Scotland. A Report to the National Goose Forum. RSPB/BASC.

Westmacott B., 2000. Estimation of Damage caused by Geese in Kintyre. *BSc dissertation (unpublished) University of Edinburgh.*

ANNEX ONE   PROJECT TEAM AND RESPONSIBILITES

Dr Douglas Macmillan (University of Aberdeen). Overall co-ordinator of the study and report editor. Responsible for Market Stall research and Tourism Focus Group analysis. Joint responsibility with Nick Hanley for the design of CE and CV studies.

Professor Nick Hanley (University of Glasgow). Main responsibilities were the analysis and reporting of benefit estimates from the CV and CE data sets. Joint responsibility with Douglas Macmillan for the design of CE and CV studies.

Professor Robert Wright (University of Stirling) provided the main statistical analysis of the CE datatsets

Dr Julie-Ann Gustanski (University of Edinburgh) was moderator of the Focus Group research involving members of the public and undertook the analysis of the discussions.

Dr Lorna Philip (University of Aberdeen) assisted with the Market Stall Research and was responsible for the Feedback Group sessions.

Mr Mike Daw and Mrs Doreen Daw (University of Aberdeen) were responsible for the agricultural cost study which included the Farmer Focus Groups, the literature review and the farm survey.

Mr Ian Patterson (University of Aberdeen), was responsible for evaluating and advising on the key scientific information on wild geese in Scotland.

Guy Garrod (University of Newcastle-upon-Tyne) acted as independent expert scrutineer of the valuation study

System Three was responsible for convening and hosting the focus group/feed back sessions and for implementing the face to face CV and CE surveys.

Collaborators
The RSPB provided full access to their reserves at Strathbeg and on Islay and provided reports and other information on wild geese in Scotland.

CRU RESEARCH - PUBLICATIONS LIST FROM 2000

Drug Misuse in Scotland: Simon Anderson & Martin Frischer. (2000) (£5.00)
Summary avialable: Crime and Criminal Justice Research Findings No.17

Intermediate Diets, First Diets and Agreement of Evidence in Criminal Cases: An Evaluation: Frazer McCallum & Professor Peter Duff (Aberdeen University Faculty of Law). (2000) (£5.00)
Summary available: Crime and Criminal Justice Research Findings No.42

The Experience of Violence and Harrasment of Gay Men in the City of Edinburgh: Colin Morrision & Andrew Mackay (The TASC Agency). (2000) (£5.00)
Summary available: Crime and Criminal Justice Research Findings No.41

The Development of the Scottish Partnership on Domestic Abuse and Recent Work in Scotland: Dr Shelia Henderson (Reid Howie Associates). (2000) (£5.00)

Children, Young People and Crime in Britain and Ireland: From Exclusion to Inclusion –1998 Conference Papers: Monica Barry (University of Stirling), Joe Connolly (Action for Children), Olwyn Burke, Dr J Curran (Central Research Unit, Scottish Executive). (2000) (£5.000)

Overview of Written Evidence Received as Part of the Review of the Public Health Function in Scotland :
Summary available only: General Research Finding No.4

Assessment of the Voter Education Campaign for the Scottish Parliament Elections: (Scotland Office Publication): Andra Laird, Sue Granville & Jo Fawcett (George Street Research). (2000) (£5.00)

Transport Provision for Disabled People in Scotland: Sheila and Brian Henderson, Reid-Howie Associates. (1999) (£10.00)
Summary available: Development Department Research Findings No.76

Transport Provision for Disabled People in Scotland: Summary: Sheila and Brian Henderson, Reid-Howie Associates. (1999) (£5.00)

Review of the Experience of Community Councils as Statutory Consultees on Planning Applications: Ewan McCraig, MVA. (2000) (£5.00)
Summary available Development Department Research Findings No.77

Family Support and Community Care: A Study of South Asian Older People: Alison Bowes and Naira Dar with the assistance of Archana Srivastava (University of Stirling). (2000) (£6.00)

An Evaluation of the SACRO (Fife) Young Offender Mediation Project: Becki Sawyer, System 3. (2000) (£5.00)
Summary available: Crime and Criminal Justice Research Findings No.43

Development Department Research 2000-2001: (2000) (Free)

Environment Group Research Programme 2000-2001: (2000) (Free)

Social Inclusion Bulletin No.3: (2000) (Free)

Road Safety Education in the Scottish Curriculum: Tony Graham, ODS Ltd. (2000) (£5.00)
Summary available: Development Department Research Findings No.78

The Role of Information and Communications Technology in Road Safety Education: BITER – The British Institute of Traffic Education Research. (2000) (£5.00)
Summary available: Development Department Research Findings No.79

Road Accidents and Children Living in Disadvantaged Areas: A Literature Review: David White, Robert Raeside and Derek Barker. (2000) (£5.00)
Summary available: Development Department Research Findings No.81

Evaluation of Scottish Road Safety Campaign Travel Packs: Sharon Reid, Andra Laird & Jo Fawcett. (2000) (£5.00)
Summary available: Development Department Research Findings No.82

Audit of ICT Initiatives: In Social Inclusion Partnerships and Working for Communities Pathfinders in Scotland: Joanna Gilliatt, Doug Maclean & Jenny Brogden, Lambda Research & Consultancy Ltd. (2000) (£5.00)

Researching Ethnic Minorities in Scotland: Reid-Howie Associates. (2000) (Free)

A Comparative Evaluation of Greenways and Conventional Bus Lanes: Colin Buchanan and Partners. (2000) (£5.00)
Summary available: Development Department Research Findings No.83

Advertising Planning Proposals: James Barr Planning Consultants. (2000) (£5.00)
Summary available: Development Department Research Findings No.84

The Role of Pre-Application Discussions and Guidance in Planning: Peter Gibson and Robert Stevenson, The Customer Managerment Consultancy Ltd. (2000) (£5.00)
Summary available: Development Department Research Findings No.85

Developing Markets for Recyclable Materials in Scotland: Prioritising Materials: Enviros RIS Ltd in association with Clean Washington Centre. (2000) (Free).
Summary only available: Environment Group Research Findings No.6

The Development of the Scottish Partnership on Domestic Abuse and recent Work in Scotland: Dr S Henderson, Reid Howie Associates. (2000) (£5.00).

Evaluation of the Airborne Initiative (Scotland): Gill McIvor, Vernon Gayle, Kirstina Moodie, Stirling University and Ann Netten, University of Kent. (2000) (£5.00)
Summary available: Crime and Criminal Justice Research Finding No.45

A Review of the Research Literature on Serious and Sexual Offenders: Clare Connelly and Shanti Williamson, University of Glasgow. (2000) (£8.00).
Summary available: Crime and Criminal Justice Research Finding No.46

The Quality of Services in Rural Scotland: Steven Hope, Simon Anderson and Becki Sawyer, System Three. (2000) (£10.00).
Summary available: Rural Affairs Research Findings No.5.

Social Exclusion in Rural Areas; A Literature Review and Conceptual Framework: Mark Shuckshank and Lorna Philip, University of Aberdeen. (2000) (£10.00).
Summary available: Rural Affairs Research Findings No.6.

Charities Report:
1. Scottish Charity Legislation; Full Report (2000) (£15.00).
2. Scottish Charity Legislation; Executive Summary (2000) (Free).
3. Scottish Charity Legislation; Annexes (2000) (£5.00).
4. Public Charitable Collections (2000) (£5.00).
5. Public Trusts and Educational Endowments (2000) (£5.00).
University of Dundee.
Summary available: Legal Studies Research Finding 26

Meeting in the Middle: A Study of Solicitors' and Mediators Divorce Practice: Fiona Myers and Fran Wasoff. University of Edinburgh. (2000) (£5.00)
Summary available: Legal Studies Research Finding No.25.

Real Burdens: Survey of Owner Occupiers' Understanding of Title Conditions: Andra Laird and Emma Peden, George Street Research (2000) (£5.00)
Summary available: Legal Studies Research Finding No.27

Survey of Complainers to the Scottish Legal Services Ombudsman: The Customer Management Consultancy Ltd (2000) (£5.00)

An Evaluation of Electronically Monitored Restriction of Liberty Orders: David Lobley and David Smith, Lancaster University. (2000) (£5.00)
Summary available: Crime and Criminal Justice Research Finding No.47

Interviewing and Drug Testing of Arrestees in Scotland: A Pilot Study of the Arrestee Drug Abuse Monitoring (ADAM) Methodology: Neil McKeganey, Clare Connelly, Lesley Reid & John Norrie University of Glasgow, Janusz Knepil Gartnavel General Hospital Glasgow. (2000) (£5.00)
Summary available: Crime and Criminal Justice Research Finding No 48

The Role of Sport in Regenerating Deprived Urban Areas: Fred Coalter with Mary Allison and John Taylor, Centre for Lesisure Research. (2000) (£5.00)
Summary available: Development Department Research Finding No 86

"Huts and Hutters" in Scotland: Research Consultancy Services. (2000) (£5.00)

Motivations to Public Service: Sue Granville and Andra Laird, George Street Research. (2000) (Free)
Summary only available: Development Department Research Findings No.87

The What, Where and When of Being a Councillor: Paolo Vestri and Stephen Fitzpatrick, Scottish Local Government Information Unit. (2000) (Free)
Summary only available: Development Department Research Findings No.88

Future Patterns of Retailing in Scotland: John Dawson (Professor of Marketing, The University of Edinburgh and Visiting Professor ESADE, Barcelona). (2000) (£5.00)
Summary available: Development Department Research Findings No.91

Women's Issues in Local Partnership Working: Gill Scott, Gill Long, Usha Brown, Jane McKenzie, Scottish Poverty Information Unit). (2000) (£5.00)
Summary available: Social Inclusion Research Findings No.1

Moving On: A Survey of Travellers' Views: Delia Lomax, Sharon Lancaster (School of Planning & Housing, ECA/Heriot-Watt University) and Patrick Gray (Housing Research Centre, Magee College, University of Ulster). (2000) (£5.00)
Summary available: Development Department Research Findings No.94

Monitoring the Children (Scotland) Act 1995: Pilot Study: Jeremy Hardin, MVA; Professor Alastair Bissett-Johnson, University of Dundee; Shona Main, University of Dundee. (2000) (£5.00)

Accessibility: Review of Measuring Techniques and their Application: Derek Halden; David McGuigan; Andrew Nisbet and Alan McKinnon (Derek Halden Consultancy). (2000) (£5.00)
Summary available: Development Department Research Findings No.89

Research into the Basis for Local and National Estimates of the Number of BTS Houses in Scotland: DTZ Pieda Consulting in association with Dr J I Ansell. (2000) (£5.00)
Summary available: Development Department Research Findings No.90

Pedestrian Perceptions of Road Crossing Facilities: J M Sharples and J P Fletcher. (2000) (£5.00)
Summary available: Development Department Research Findings No.92

Public Attitudes Towards Wind Farms in Scotland: Anna Dudleston, System Three Social Research. (2000) (£5.00)
Summary available: Development Department Research Findings No.93

Social Inclusion Bulletin No.4: (2000) (Free)

Research on Repeat Victimisation in Scotland: Mandy Shaw and Ken Pease, Applied Criminology Group, University of Huddersfield (2000) (£5.00)
Summary available: Crime & Criminal Justice Research Findings No.44

Social Work and Criminal Justice: The Longer Term Impact of Supervision: Gill McIvor and Monica Barry, Social Work Research Centre, University of Stirling (2000) (£5.00)
Summary available: Crime and Criminal Justice Research Findings No.50

An Evaluation of the Parent Information Programme: Gillian Mayes, John Gilles, Greame Wilson, University of Glasgow, Raymond Macdonald, Glasgow Caledonian University (2000) (£5.00)
Summary available: Legal Studies Research Findings No.29

The Use of Civil Legal Remedies for Neighbour Nuisance in Scotland: Rowland Atkinson, Tom Mullen, and Suzie Scott, University of Glasgow (2000) (£8.00)
Summary available: Legal Studies Research Findings No 28

Results of the Scottish Executive Staff Survey 2000: Patrick Barron and Andrew Fleming. (2000) (Free)
Summary only available: General Research Research Findings No.5

The 2000 Scottish Crime Survey: First Results: MVA Ltd. (2000) (Free)
Summary only available: Crime and Criminal Justice Research Findings No.51

Postal Witness Citation and Countermanding: An Evaluation of the Mechanised System Piloted in Glasgow, Ayr and Lanark: Ian Clark. (2000) (Free)
Summary only available: Crime and Criminal Justice Research Findings No.49

Solicitor Advocates in Scotland: A Statistical Analysis: Debbie Headrick. (2000) (£5.00)
Summary available: Legal Studies Research Findings No.32

Solicitor Advocates in Scotland: The Impact on Clients: Gerard Hanlon and John Jackson, (2000) (£5.00)
Summary available: Legal Studies Research Findings No.33

Solicitor Advocates in Scotland: The Impact on the Legal Profession: Karen Kerner. (2000) (Free)
Summary only available: Legal Studies Research Findings No.34

Solicitor Advocates in Scotland: A Research Overview: Alison Platts. (2000) (£5.00)
Summary available: Legal Studies Research Findings No.35

Green Commuter Plans: Do They Work?: Napier University, Transport Research Institute. (2000) (£5.00)
Summary available: Development Department Research Findings No.95

The Role of the Arts in Regeneration: Alan Kay and Glenys Watt, Blake Stevenson Ltd. (2000) (£5.00)
Summary available: Development Department Research Findings No.96

Assessment of Innovative Approaches to Testing Community Opinion: George Street Research Ltd. (2000) (£5.00)
Summary available: Social Inclusion Research Findings No.2

Local Authority Waste Management Costs Study: Enviros Aspinwall. (2000) (£5.00)
Summary available: Environment Group Research Findings No.7

Credit Union Development Activity in Scotland: Centre for Economic Development and Area Regeneration (CEDAR) and The Planning Exchange. (2000) (£5.00)

Using People's Juries in Social Inclusion Partnerships: Guidance for SIPs: Jo Fawcett, Sue Granville and Andra Laird (George Street Research). (2000) (Free)

The People's Panel in Scotland's Social Inclusion Partnerships: Guidance for SIPs: Robin Clarke, Ruth Rennie, Clare Delap and Vicki Coombe. (2000) (Free)
Summary available: Development Department Research Findings No.97

Women and Transport: Moving Forward: Reid-Howie Associates. (2000) (£5.00)
Summary available: Development Department Research Findings No.98

An Evaluation of Section 18 of the Mental Health (Scotland) Act 1984: Alison Bean, Ann McGuckin and Suzi Macpherson; Legal Studies Research Branch-Central Research Unit. (2001) (£5.00)
Summary available: Legal Studies Research Findings No. 30

An Evaluation of Guardianship Under the Mental Health (Scotland) Act 1984: Alison Bean, Ann McGuckin and Suzi Macpherson; Legal Studies Research Branch- Central Research Unit. (2001) (£5.00)
Summary available: Legal Studies Research Findings No. 31

Review of the Literature Relating to the Mental Health Legislation.

The Criminal Histories of 372 Suspected Drug Offenders: Inspector Joe McGallagly and Sergeant Brian Dunn – Strathclyde Police Policy and Development Branch. (2001) (Free)
Summary only available: Crime and Criminal Justice Research Findings No. 52

Family Mediation Services for Minority Ethnic Communities in Scotland: Vibha Pankaj, Family Mediation Scotland. (2001) (£5.00)
Summary available: Legal Studies Research Findings No. 36

Evaluation of the Scottish Cycle Challenge Initiative: Derek Halden and David McGuigan, (Derek Halden Consultancy) and Jonathan Toy, (Ove Arup and Partners). (2001) (£5.00)
Summary available: Development Department Research Findings No. 100

Evaluation of the Local Rural Partnership Scheme: Louise Brown Research. (2001) (£5.00)
Summary available: Agricultural Policy Coordination and Rural Development Research Findings No. 7.

Public Assistance to Rural Land in Scotland: Richard Cowell, Gillian Bristow, Terry Marsden and Alex Franklin (Cardiff University). (2001) (£5.00)
Summary available: Agricultural Policy Coordination and Rural Development Research Findings No. 8

Scottish Waste Statistics for 1997 and 1998: Environmental Resources Management. (2000) (£5.00)
Summary available: Environment Group Research Findings No. 9

Review of the Environmental Effects of the Landfill Tax in Scotland: EAG Environ. (2000) (£5.00)
Summary available: Environment Group Research Findings No. 10

Obsolete Commercial and Industrial Buildings: EKOS Limited and Ryden Property Consultants. (2001) (£5.00)
Summary available: Development Departement Research Findings No. 101

4

An Exploration of Regional Climate Change Scenarios for Scotland: Mike Hulme and Xianfu Lu (University of East Anglia). (2001) (£5.00)
Summary available: Environment Group Research Findings No. 11

Climate Change: North Atlantic Comparisons: Andrew Kerr and Simon Allen (University of Edinburgh). (2000) (£5.00)
Summary available: Environment Group Research Findings No. 8

Recycled Aggregates in Scotland: M G Winter (TRL Ltd.) and C Henderson (ERM Ltd). (2001) (£5.00)
Summary available: Development Department Research Findings No. 103

Evaluation of the Empty Homes Initiative: Caledonian Economics Ltd, in association with Arneil Johnston. (2001) (£5.00)
Summary available: Development Department Research Findings No. 105

Freagarrach: An Evaluation of a Project for Persistent Juvenile Offenders: David Lobley, David Smith and Christina Stern (Lancaster University). (2001) (£5.00)

Social Work and Criminal Justice: The Longer Term Impact of Supervision: Gill McIvor and Monica Barry (University of Stirling). (2000) (£5.00)

A Review of Literature Relating to the Mental Health Legislation: Dr Jacqueline Atkinson, Lesley Patterson (University of Glasgow). (2001) (£5.00)

Recreational Drugs and Driving: Prevalence Survey: Dave Ingram, Becki Lancaster and Steven Hope (System Three Social Research). (2001) (£5.00)
Summary available: Development Department Research Findings No. 102

Recreational Drug Use and Driving: A Qualitative Study: Joanne Neale, Neil McKeganey, Gordon Hay, (Centre for Drug Use Research – University of Glasgow), and John Oliver, (Department of Forensic Medicine, University of Glasgow). (2001) (£5.00)
Summary available: Development Department Research Findings No. 102

People's Juries in Social Inclusion Partnerships: Six Month Follow-Up of the Jury Process: (Ruth Rennie – Office for Public Management). (2001) (Free)
Summary only available: Development Department Research Findings No.106

The Role of the Planning System in the Provision of Housing: School of Planning and Housing (Edinburgh College of Art) in association with the Department of Building Engineering and Surveying, (Heriot Watt University). (2001) (£5.00)
Summary available: Development Department Research Findings No.107

Interchange and Travel Choice: Volume 1: Mark Wardman, Julian Hine and Stephen Stradling. (2001) (£5.00)
Summary available: Development Department Research Findings No.99

Interchange and Travel Choice: Volume 2: Mark Wardman, Julian Hine and Stephen Stradling. (2001) (£5.00)
Summary available: Development Department Research Findings No.99

The Influence of Local Biodiversity Action Plans in Unitary Authority LA21 Process and Community Planning: EKOS Limited. (2001) (£5.00)
Summary available: Countryside and Natural Heritage Research Finidngs No.4

Social Inclusion Partnerships and Locality Budgeting: Professor Michael Carley. (2001) (£5.00)
Summary available: Development Department Research Findings No.109

Social Inclusion Research Bulletin No.5: (2001) (Free)

Development Department Research 2001-2002: (2001) (Free)

Environment Group Research Programme 2001-2002: (2001) (Free)

The Role of Transport in Social Exclusion in Urban Scotland: Julian Hine and Fiona Mitchell (Transport Research Institute, Napier University). (2001) (£5.00)
Summary available: Development Department Research Findings No.110

The Role of Transport in Social Exclusion in Urban Scotland: Literature Review: P Gaffon, J P Hine and F Mitchell (Transport Research Institute, Napier University). (2001) (£5.00)

Climate Change and Changing Snowfall Patterns in Scotland: Dr John Harrison & Dr Sandy Winterbottom (University of Stirling) and Dr Richard Johnson (Mountain Environments, Callander). (2001) (£5.00)
Summary available: Environment Group Research Findings No.14

Climate Change: Review of Levels of Protection Offered By Flood Prevention Schemes: D J Price and G McInally (Babtie Group). (2001) (£5.00)
Summary available: Environment Group Research Findings No.12

Good Practice in Housing Management: Case Studies, Conclusions and Recommendations: Suzie Scott, Hector Currie, Jo Dean and Keith Kintrea. (2001) (£5.00)
Summary available: Development Department Research Findings No.112

Good Practice in Housing Management: A Review of the Literature: Suzie Scott (Editor), Hector Currie, Suzanne Fitzpatrick, Hal Pawson, Keith Kintrea, Ann Rosengard and Jenny Tate. (2001) (£5.00)
Summary available: Development Department Research Findings No.112

Good Practice in Housing Management: A Review of Progress: Suzie Scott, Hector Currie, Suzanne Fitzpatrick, Margaret Keoghan, Keith Kintrea, Hal Pawson and Jenny Tate. (2001) (£5.00)
Summary available: Development Department Research Findings No.112

Impact of Agricultural Practices and Catchment Characteristics on Ayrshire Bathing Waters: Mark Aitken, David W Merrilees and Alistair Duncan (Environment Division, SAC, Auchincruive). (2001) (£5.00)

20mph Speed Reduction Initiative: Archie Burns, Neil Johnstone and Neil MacDonald, Halcrow Group Ltd. (2001) (£10.00)
Summary available: Development Department Research Findings No.104

Rethinking Open Space - Open Space Provision and Management: A Way Forward: Kit Campbell Associates, Edinburgh. (2001) (£5.00)
Summary available: Development Department Research Findings No.108

Evaluation of a New Deal for Young People in Scotland - Phase1: Jane Lakey, Genevieve Knight. (2001) (£5.00)
Summary available: Enterprise and Lifelong Learning Department Research Findings No.1

Sustainability Indicators for Waste, Energy and Travel for Scotland: Entec UK Ltd. (2001) (£5.00)
Summary available: Environment Group Research Findings No.13

The Role of Scottish Local Initiatives in Implementing the Principles of Integrated Coastal Zone Management: Susan Gubbay. (2001) (£5.00)
Summary available: Countyside and Natural Heritage Research Findings No.5

A Coastal Management Trust for Scotland: John Firn and Derek McGlashan (Firn Crichton Roberts Ltd). (2001) (£5.00)
Summary available: Countryside and Natural Heritage Research Findings No.6

Further information on any of the above is available by contacting:

Chief Research Officer
Scottish Executive Central Research Unit
Room J1-5
Saughton House
Broomhouse Drive
Edinburgh
EH11 3XA

or by accessing the World Wide Website: www.scotland.gov.uk